Lecture Notes
in Business Information Processing
414

Series Editors

Wil van der Aalst ⓘ
RWTH Aachen University, Aachen, Germany
John Mylopoulos ⓘ
University of Trento, Trento, Italy
Michael Rosemann ⓘ
Queensland University of Technology, Brisbane, QLD, Australia
Michael J. Shaw
University of Illinois, Urbana-Champaign, IL, USA
Clemens Szyperski
Microsoft Research, Redmond, WA, USA

More information about this series at http://www.springer.com/series/7911

Uchitha Jayawickrama ·
Pavlos Delias · María Teresa Escobar ·
Jason Papathanasiou (Eds.)

Decision Support Systems XI

Decision Support Systems, Analytics and Technologies in Response to Global Crisis Management

7th International Conference on Decision
Support System Technology, ICDSST 2021
Loughborough, UK, May 26–28, 2021
Proceedings

 Springer

Editors
Uchitha Jayawickrama 🆔
Loughborough University
Loughborough, UK

María Teresa Escobar 🆔
University of Zaragoza
Zaragoza, Spain

Pavlos Delias 🆔
International Hellenic University
Kavala, Greece

Jason Papathanasiou 🆔
University of Macedonia
Thessaloniki, Greece

ISSN 1865-1348   ISSN 1865-1356 (electronic)
Lecture Notes in Business Information Processing
ISBN 978-3-030-73975-1   ISBN 978-3-030-73976-8 (eBook)
https://doi.org/10.1007/978-3-030-73976-8

This Springer imprint is published by the registered company Springer Nature Switzerland AG
The registered company address is: Gewerbestrasse 11, 6330 Cham, Switzerland

Preface

This eleventh edition of the EURO Working Group on Decision Support Systems published in the LNBIP series presents a selection of full papers from the seventh International Conference on Decision Support System Technology (ICDSST 2021), initially planned to be held in Loughborough, UK, during May 26–28, 2021, but eventually held virtually due to the ongoing COVID-19 pandemic. The conference's main theme was "Decision Support Systems, Analytics and Technologies in response to Global Crisis Management" and the principal aim was to investigate the role Decision Support Systems (DSS) and related technologies can play in both mitigating the impact of pandemics and post-crisis recovery.

The Euro Working Group on Decision Support Systems (EWG-DSS) planned this conference series of International Conference on Decision Support System Technology (ICDSST), starting with ICDSST 2015 in Belgrade, to consolidate the tradition of annual events organized by the EWG-DSS in offering a platform for European and international DSS communities, comprising the academic and industrial sectors, to present state-of-the-art DSS research and developments, to discuss current challenges that surround decision-making processes, to exchange ideas about realistic and innovative solutions, and to co-develop potential business opportunities. Building on this tradition, ICDSST 2021 included the following scientific topic areas:

- Decision Support Systems: Advances and Future Trends
- Multi-Attribute and Multi-Criteria Decision Making
- Knowledge Management, Acquisition, Extraction, Visualization and Decision Making
- Multi-Actor Decision Making: Group and Negotiated Decision Making
- Collaborative Decision Making and Decision Tools
- Discursive and Collaborative Decision Support Systems
- Mobile and Cloud Decision Support Systems
- GIS and Spatial Decision Support Systems
- Data Science, Data Mining, Text Mining, and Sentimental Analysis
- Big Data Analytics
- Imaging Science (Image Processing, Computer Vision and Pattern Recognition)
- Human-Computer Interaction
- Internet of Things
- Social Network Analysis for Decision Making
- Simulation Models and Systems, Regional Planning, Logistics and SCM
- Business Intelligence, Enterprise Systems and Quantum Economy
- Machine Learning, Natural Language Processing, Artificial Intelligence
- Virtual and Augmented Reality
- New Methods and Technologies for Global Crisis Management
- Analytics for Mitigating the Impact of Pandemics

- Intelligent DSS for Crisis Prevention
- Innovative Decision Making during Global Crises
- New DSS Approaches for Post-Crisis Economic Recovery
- Decision Making in Modern Education
- Decision Support Systems for Sports
- General DSS Case Studies (Education, E-Government, Energy, Entrepreneurship, Environment, Healthcare, Industrial Diversification and Sustainability, Innovation, Logistics, Natural Resources, etc.)

These topics reflect some of the essential areas of study within Decision Support Systems, as well as the research interests of the group members. This rich variety of themes, advertised not only to the (more than three hundred) members of the group but also to a broader audience, allowed us to gather contributions regarding the implementation of decision support processes, methods, and technologies in a large variety of domains. Hence, this EWG-DSS LNBIP Springer edition collates high-quality contributions of full papers, which were selected through single-blind peer review. At least two members of the Program Committee reviewed each submission in the first part of a rigorous two-stage process. The second stage involved the volume editors judging whether the revised versions did indeed address the issues that reviewers had raised. Papers that didn't address properly all of the issues were either accepted to the conference but not included in this volume, or were not accepted at all. Finally, we selected 10 out of 44 submissions, which corresponds to a 22.7% acceptance rate, to be included in this eleventh edition of Decision Support Systems.

We proudly present the selected contributions, organized in two sections:

1. *Multiple Criteria Approaches*. This section concerns methods and applications of Multiple Criteria Decision Aid (MCDA), including specialized software development. First, He Huang, Koen Mommens, Philippe Lebeau, and Cathy Macharis present "The Multi-Actor Multi-Criteria Analysis (MAMCA) for Mass-Participation Decision Making", a tool that allows for the involvement of multiple stakeholders within a decision-making process, and for various relevant checks like the homogeneity and heterogeneity control of the stakeholders per group. Then, "Using FITradeoff Method for supply selection with decomposition and holistic evaluations for Preference Modelling" by Lucia Reis, Peixoto Roselli, and Adiel Teixeira de Almeida deals with additive aggregation in the context of Multi-Attribute Value Theory and discusses the combination of two perspectives of preference modelling in the FITradeoff method for a supply selection decision problem, while delivering a DSS that provides graphical and tabular visualizations. This section continues with "DEX2Web – A Web-Based Software Implementing the Multiple-Criteria Decision-Making Method DEX" by Adem Kikaj and Marko Bohanec, a specialized online suite of tools for the application of the popular method DEX to help individuals and groups with their decision-making challenges. DEX and other MCDA methods are compared over the decision problem of employee selection in the next article, "Comparison of AHP, PAPRICA, PRO-METHEE, DEX and TOPSIS on an application for employee selection" by Anton Stipec and Biljana Mileva Boshkoska. The section closes with "A Survey on

Criteria for Smart Home Systems with Integration into the Analytic Hierarchy Process" by Georg Wieland and Herwig Zeiner, where the popular AHP methodology is used through a survey on the most important criteria relevant to smart home systems, discussing topics like the usability, the sustainability and the complexity of smart home applications.

2. *Advances in Decision Support Systems' technologies and methods.* This section focuses on methods, techniques, approaches, and technologies that advance the research of the DSS field. Jinyi Liu and Patrick Stacey present "Modelling the Effects of Lockdown and Social Distancing in the Management of the Global Coronavirus Crisis - Why the UK Tier System Failed" where a variety of lockdown circumstances were simulated to examine the importance of social distancing, lockdown, and quarantine measures. Simulation was also used as a tool for decision making in an industrial context in the next article, "A case study initiating discrete event simulation as a tool for decision making in I4.0 manufacturing" by Kristina Eriksson and Ted Hendberg. Their approach demonstrates the power of analytics to handle the uncertainty by infusing responsiveness and flexibility into the modelled systems. Then, Tatiana Levashova, Alexander Smirnov, Andrew Ponomarev, and Nikolay Shilov present their work on "Methodology for Multi-Aspect Ontologies Development: Use Case of DSS Based on Human-Machine Collective Intelligence". Focusing on ontology development methodologies they outline the scope of a multi-aspect ontology application and introduce a collective intelligence environment for decision support. The section continues with a set of works relevant to Artificial Intelligence (AI). In "Investigating oversampling techniques for fair machine learning models" Sanja Rančić, Sandro Radovanović, and Boris Delibašić are challenged by the probable unethical and illegal consequences of AI applications and propose oversampling techniques to increase fairness, without decreasing in predictive accuracy of the methods. Again on AI, Ralph Grothmann and Ulrike Dowie present "Using AI to Advance Factory Planning: A Case Study to Identify Success Factors of Implementing an AI-Based Demand Planning Solution". The focus is on constructing a forecasting model for customer demand and on joining the forecasts with uncertainty measures to support the decisions of the demand planning department under uncertainty.

We would like to thank the many people who contributed majorly to the success of this LNBIP book. First of all, we would like to thank Springer for providing us with the opportunity to guest edit this DSS volume, and we wish to express our sincere gratitude to Ralf Gerstner and Christine Reiss, who dedicated their time to guide and advise us during the volume editing process. Secondly, we need to thank all the authors for submitting their state-of-the-art work for consideration to this volume, managing to overcome all the obstacles that have affected scholars around the globe since the pandemic began. From our point of view, ICDSST 2021 was yet another confirmation that the DSS community is vivid, active, and has a great potential for contributions to science. It really gives us courage and stimulates us to continue the series of International Conferences on Decision Support System Technology. Finally, we express our deep gratitude to the reviewers, members of the Program Committee, who volunteered to assist in the improvement and the selection of papers, under (to be honest) a tight schedule. We believe that this EWG-DSS Springer LNBIP volume presents a rigorous selection of high-quality papers addressing the conference theme. We hope that readers will enjoy the publication!

March 2021

Uchitha Jayawickrama
Pavlos Delias
María Teresa Escobar
Jason Papathanasiou

Organization

Conference Chairs

Shaofeng Liu   University of Plymouth, UK
Uchitha Jayawickrama   Loughborough University, UK

Steering Committee – EWG-DSS Coordination Board

Shaofeng Liu   University of Plymouth, UK
Boris Delibašić   University of Belgrade, Serbia
Jason Papathanasiou   University of Macedonia, Greece
Isabelle Linden   University of Namur, Belgium
Pavlos Delias   International Hellenic University, Greece

Program Committee

Adiel Teixeira de Almeida   Federal University of Pernambuco, Brazil
Alex Duffy   University of Strathclyde, UK
Alex Zarifis   Loughborough University, UK
Alexander Smirnov   Russian Academy of Sciences, Russia
Alexis Tsoukias   University Paris Dauphine, France
Alok Choudhary   Loughborough University, UK
Ana Paula Cabral   Federal University of Pernambuco, Brazil
Andy Wong   University of Strathclyde, UK
Ben C. K. Ngan   Worcester Polytechnic Institute, USA
Bertrand Mareschal   Université Libre de Bruxelles, Belgium
Boris Delibašić   University of Belgrade, Serbia
Carlos Henggeler Antunes   University of Coimbra, Portugal
Christian Colot   University of Namur, Belgium
Daouda Kamissoko   University of Toulouse, France
Dragana Bečejski-Vujaklija   Serbian Society for Informatics, Serbia
Emilio Larrodé   Zaragoza University, Spain
Fátima Dargam   SimTech Simulation Technology/ILTC, Austria
Femi Olan   Northumbria University, UK
Fernando Tricas   Zaragoza University, Spain
Francisco Antunes   Beira Interior University, Portugal
François Pinet   Cemagref/Irstea, France
Gloria Philipps-Wren   Loyola University Maryland, USA
Guy Camilleri   Toulouse III University/IRIT, France
Hing Kai Chan   University of Nottingham Ningbo China, UK/China
Isabelle Linden   University of Namur, Belgium
Jan Mares   University of Chemical Technology, Czech Republic

Jason Papathanasiou	University of Macedonia, Greece
Jean-Marie Jacquet	University of Namur, Belgium
João Lourenço	Universidade de Lisboa, Portugal
João Paulo Costa	University of Coimbra, Portugal
Jorge Freire de Souza	Engineering University of Porto, Portugal
José Maria Moreno Jimenez	Zaragoza University, Spain
Kayode Odusanya	Loughborough University, UK
Konstantina Spanaki	Loughborough University, UK
Konstantinos Vergidis	University of Macedonia, Greece
Maduka Subasinghage	Auckland University of Technology, New Zealand
Marc Kilgour	Wilfrid Laurier University, Canada
María Teresa Escobar	Zaragoza University, Spain
Marko Bohanec	Jozef Stefan Institute, Slovenia
Md Asaduzzaman	Staffordshire University, UK
Nikolaos Matsatsinis	Technical University of Crete, Greece
Nikolaos Ploskas	University of Macedonia, Greece
Panagiota Digkoglou	University of Macedonia, Greece
Pascale Zaraté	IRIT/Toulouse University, France
Pavlos Delias	International Hellenic University, Greece
Peter Kawalek	Loughborough University, UK
Rudolf Vetschera	University of Vienna, Austria
Sandro Radovanović	University of Belgrade, Serbia
Sean Eom	Southeast Missouri State University, USA
Shaofeng Liu	University of Plymouth, UK
Stefanos Tsiaras	Aristotle University of Thessaloniki, Greece
Uchitha Jayawickrama	Loughborough University, UK
Wim Vanhoof	University of Namur, Belgium

Local Organizing Team

Peter Kawalek	Loughborough University, UK
Kayode Odusanya	Loughborough University, UK
Konstantina Spanaki	Loughborough University, UK
Crispin Coombs	Loughborough University, UK

Sponsors

Working Group on Decision Support Systems
(https://ewgdss.wordpress.com)

Association of European Operational Research Societies
(www.euro-online.org)

Institutional Sponsors

Loughborough University
(http://www.lboro.ac.uk)

School of Business and Economics (SBE)
(http://www.lboro.ac.uk/departments/sbe/)

Centre for Information Management (CIM)
(http://www.lboro.ac.uk/departments/sbe/cim/)

Graduate School of Management, Faculty of Business,
University of Plymouth, UK
(http://www.plymouth.ac.uk/)

Faculty of Organisational Sciences,
University of Belgrade, Serbia
(http://www.fon.bg.ac.rs/eng/)

University of Namur, Belgium
(http://www.unamur.be/)

University of Macedonia, Department of Business
Administration, Thessaloniki, Greece
(http://www.uom.gr/index.php?newlang=eng)

International Hellenic University, Greece
(https://www.ihu.gr/en/enhome)

Federal University of Pernambuco, Brazil
(https://www.ufpe.br/inicio)

Contents

Multiple Criteria Approaches

The Multi-Actor Multi-Criteria Analysis (MAMCA) for Mass-Participation
Decision Making. 3
   He Huang, Koen Mommens, Philippe Lebeau, and Cathy Macharis

Using FITradeoff Method for Supply Selection with Decomposition
and Holistic Evaluations for Preference Modelling. 18
   Lucia Reis Peixoto Roselli and Adiel Teixeira de Almeida

DEX2Web – A Web-Based Software Implementing the Multiple-Criteria
Decision-Making Method DEX. 30
   Adem Kikaj and Marko Bohanec

Comparison of AHP, PAPRICA, PROMETHEE, DEX and TOPSIS
on an Application for Employee Selection 44
   Anton Stipeč and Biljana Mileva Boshkoska

A Survey on Criteria for Smart Home Systems with Integration
into the Analytic Hierarchy Process. 55
   Georg Wieland and Herwig Zeiner

Advances in Decision Support Systems' Technologies and Methods

Modelling the Effects of Lockdown and Social Distancing in the
Management of the Global Coronavirus Crisis - Why the UK Tier
System Failed. 69
   Jinyi Liu and Patrick Stacey

A Case Study Initiating Discrete Event Simulation as a Tool for Decision
Making in I4.0 Manufacturing 84
   Kristina Eriksson and Ted Hendberg

Methodology for Multi-aspect Ontology Development. 97
   Alexander Smirnov, Tatiana Levashova, Andrew Ponomarev,
   and Nikolay Shilov

Investigating Oversampling Techniques for Fair Machine
Learning Models. 110
   Sanja Rančić, Sandro Radovanović, and Boris Delibašić

Using AI to Advance Factory Planning: A Case Study to Identify Success
Factors of Implementing an AI-Based Demand Planning Solution   124
   Ulrike Dowie and Ralph Grothmann

Author Index   135

Multiple Criteria Approaches

The Multi-Actor Multi-Criteria Analysis (MAMCA) for Mass-Participation Decision Making

He Huang$^{(\boxtimes)}$ ⓘ, Koen Mommens ⓘ, Philippe Lebeau ⓘ, and Cathy Macharis ⓘ

Vrije Universiteit Brussel, Boulevard de la Plaine 2, 1050 Ixelles, Belgium
He.Huang@vub.be

Abstract. The Multi-Actor Multi-Criteria Analysis is a methodology that allows for the involvement of multiple stakeholders within a decision-making process. It reveals the consensus and conflicts between the different groups of people that are involved in the evaluation but hold different interests. Nowadays, the concept of the "stakeholder" in MAMCA gradually shifts to the "stakeholder group", and there is a need for involving more than one evaluator in the stakeholder group to make sure all the voices from the group will be heard instead of being represented by one. Especially when a stakeholder group contains a large variation in interests, concerns and socio-economic characteristics. Additionally, one group can have subgroups that might be hard to reach, and therefore are not or un-der-represented in the analysis. This is typically the case for the 'citizens' stakeholder group.

In order to fulfill the needs of the involvement of many different stakeholders within stakeholder groups, the mass-participation function was developed in MAMCA and the MAMCA survey tool is designed. This tool allows the decision-maker to design the dedicated survey for the stakeholder group which needs the mass-participation function. The easy-to-understand evaluation process is used to avoid time-consuming elicitation. It is possible to check the homogeneity and heterogeneity of the stakeholders within the stakeholder group based on the socio-economic profiles collected in the survey.

Keywords: Mass participation · Multi-criteria decision making · Multi actor multi criteria analysis · Survey

1 Introduction

In the decision-making process of public management, stakeholder involvement plays an important role. The stakeholders, as individuals, have influences on the decision-making [1]. Normally they have different backgrounds, representing different organizations/groups. They have interests in the objectives of the project and will be affected by the consequence of the decision taken [2]. By involving the stakeholders, the decision-maker can have a better understanding of the objectives of the different parties, which typically leads to higher implementation acceptance and lower chances of project failure [3]. In the meantime, the stakeholders are able to voice their own interests or concerns.

© Springer Nature Switzerland AG 2021
U. Jayawickrama et al. (Eds.): ICDSST 2021, LNBIP 414, pp. 3–17, 2021.
https://doi.org/10.1007/978-3-030-73976-8_1

Furthermore, the stakeholders can be aware of the presence of other stakeholders, and the process of the evaluation can reflect their mutual interests and conflicts explicitly [4].

Multi-Actor Multi-Criteria Analysis (MAMCA) is a methodology that extends the traditional Multi-Criteria Decision-Making (MCDM) methods by allowing the inclusion of multiple stakeholders (see Fig. 1). The involvement of stakeholders in MAMCA facilitates a more rational solution in the field of energy [5], transportation [6], logistic and mobility [7].

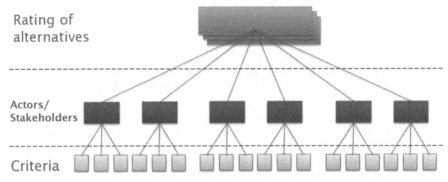

Fig. 1. MAMCA structure

In the MAMCA evaluation process, it is found that some stakeholder groups are not suitable to be represented by one or a few stakeholders. Because even when they have the same criteria, their priority to these criteria can be different [8]. Thus, a need for mass-participation comes to the table of discussion. An extended survey tool designed for mass-participation involvement in MAMCA software is developed.

In this paper, we will first explain the further developed MAMCA methodology towards a mass-participation tool. Then, the MAMCA survey tool is introduced. Finally, a didactic case study of supply chain management is applied to demonstrate the mass-participation function.

2   MAMCA Methodology Evolution

The MAMCA methodology was proposed to reach a consensus among all the stakeholders. In Fig. 2, the 7 steps of the MAMCA methodology is shown: (1) alternatives definition, (2) stakeholder analysis, (3) criteria and weights definition, (4) criteria indicators and measurement methods definition, (5) overall analysis and ranking, (6) results and (7) implementation. It is clear to see, after defining the alternatives, the stakeholder analysis is taken. Stakeholders are identified in the early stage [9]. Each stakeholder takes individual Multi-Criteria Analysis (MCA) based on his/her own criteria tree [10]. The stakeholders can evaluate the alternatives with their own preferences based on the priorities of their criteria set. They do not confront the conflicts from other stakeholders. Only at the end of the evaluation, they can check the result of their evaluation, as well as others'. In such a way, there will not be an intervention among the assessment of different stakeholders. And during the result analysis, they will be aware of the presence of other common or conflicting interests or concerns from other stakeholder groups. During the discussion of the result, the stakeholders can express their interests and explain the result of the evaluation. The decision-maker will find a win-win solution for all stakeholders easier after the discussion.

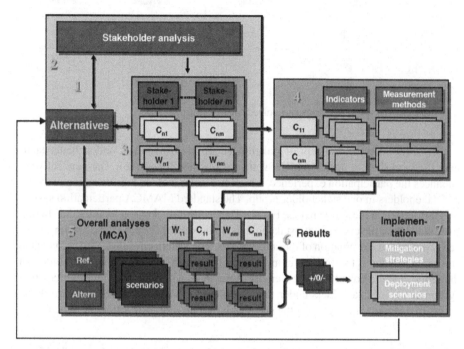

Fig. 2. MAMCA methodology [11]

After the methodology was introduced for years [12], it was found that normally there is a need for more than one stakeholder to represent their interest party. More stakeholders are invited in the workshop for the evaluation. Turcksin et al. invited 31 highly representative stakeholders from 7 different groups to assess several biofuel options for Belgium that can contribute to the binding target of 10% renewable fuels in transport by 2020 [13]. Sun et al. surveyed 48 highly representative stakeholders from 8 groups to evaluates the low-carbon transport policies in Tianjin, China [14]. Keseru et al. invited 40 participants into 7 different groups to improve mobility in the city center of Leuven, Belgium [15]. It could be foreseen that the MAMCA evaluation is not satisfied with only one representative for each group, that is, the concept of the "stakeholder" move to "stakeholder group", as it is hard for only one stakeholder to represent the whole interest and preference of his/her group. Multiple stakeholders can be invited for the evaluation of their stakeholder group. Stakeholders within one group already negotiate, but there is still a bit of struggle with loud and quiet people. They may share the same criteria, yet they can hold different priorities to the criteria (see Fig. 3).

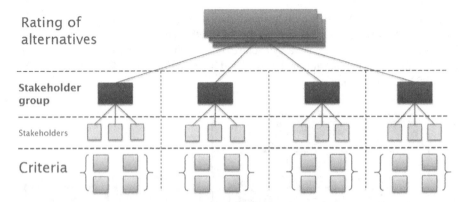

Fig. 3. Evolved MAMCA structure

To better adapt the concept of stakeholder group involvement, and to better facilitate the workshop, a new MAMCA software was developed [16]. The new software enhances the participation experience, which can better include the evaluation of multiple stakeholders in one stakeholder group. The standard MAMCA participation system was introduced in the software (see Fig. 4). The decision-maker can identify the alternatives and define the criteria with stakeholders in the workshop. And the decision-maker can coordinate the evaluation of the stakeholders. The weight allocation on criteria of the stakeholder group is the arithmetic mean of all the ranking scores of the stakeholders in the group, and the box plot of the weights' differences will be shown. This participation system can help stakeholders understand the impact on each other. They can check the points of view not only between the stakeholder groups but also within the group.

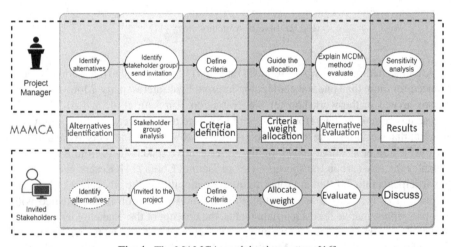

Fig. 4. The MAMCA participation system [16]

Still, for some stakeholder groups, this participation system is not well suited. Especially when there are stakeholder groups like citizens. This kind of group could have a massive amount of stakeholders, it is important to collect more profiles from the group [17]. The opinions from the group need to be heard as much as possible, as it is considered a way to reduce uncertainty and to improve the democratic legitimacy of those processes. Because the stakeholders in the group normally have different Socioeconomic status (SES), the different voices need to be heard, instead of only represented by one or limited amount during the evaluation. On the other hand, such stakeholders are hard to reach. Seeing it is always time-consuming and costly to assemble a large number of stakeholders at the same time, it is not feasible to invite all the stakeholders in the workshop for the evaluation [18]. A new evaluation model for better assessment by such stakeholder groups is needed. Thus, mass-participation decision making is proposed.

3 Mass-Participation Decision-Making in MAMCA

Mass-participation is sought targeting to certain stakeholder group, which contains the following attributes:

- A massive number of stakeholders within one stakeholder group;
- The group that requires more than one representative to voice the preferences of the group;
- The stakeholders in the group have various relevant socioeconomic status;
- The stakeholders are hard to reach and assemble;
- The stakeholders need an easy to understand and less time-consuming evaluation method.

Survey data collection is suitable for the evaluation in such a stakeholder group that fulfills the needs of the mentioned attributes [19]: Because it is not possible to gather all stakeholders in a single MAMCA workshop, the survey offers them the possibility to do the weight allocation and evaluation individually, at a non-specified time. The survey consists of the following elements: Designing and answering survey questions, weight allocation, and alternative evaluation. In the survey, the decision-maker can also ask questions on their socio-economic profiles for later research. The Profile Ranking with Order Statistics Evaluations (PROSE) is applied for the evaluation [20]. This approach combines MCDA, voting theory. After the evaluation, the decision-maker can import the survey data to the MAMCA model of the main project. It is also possible to do a post-hoc analysis to find out the homogeneity and heterogeneity within the stakeholder group. As shown in Fig. 5, the MAMCA survey model aimed for mass-participation decision-making is proposed. In such a way, the stakeholders and the decision-maker can work independently. The stakeholders can weigh the criteria and evaluate the alternatives under the assistance of the survey tool instruction, without guidance from the decision-maker, unlike the standard MAMCA participation system where the stakeholders have to participate in the physical or online workshop. In the following sub-section, the necessary steps of the model are clarified.

3.1 Designing and Answering Survey Questions

When there is a massive amount of stakeholders in one stakeholder group, instead of treating the stakeholder group as a whole all the time, there is a need to look inside

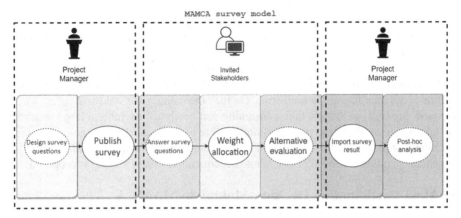

Fig. 5. MAMCA survey model

the characteristics of individuals. In a stakeholder group like citizens, the priorities and preferences of stakeholders can vary according to gender, age, income, education, etc. [21]. By collecting socio-economic profiles of the stakeholders it can provide a "bird eye view" of the stakeholder group, which helps the decision-maker identify profiles, concerns, and opinions. It displays combined and comparable statistical snapshots of the stakeholder group.

The SES are important indicators in mass-participation decision-making, as the stakeholder group like "citizens", "residents" is in a more general term, that it is possible to find a significant difference statistically of the criteria priority ranking or alternative evaluation. In that case, the stakeholders can be regrouped or divided into sub-groups [22].

The analysis of the stakeholder group's homogeneity and heterogeneity can be done by asking about some specific stakeholders' SES. The decision-maker can design survey questions for inquiring. After collecting the socio-economic profiles of the stakeholders, it is possible to do a post-hoc analysis by combining the criteria priority ranking and socio-economic profiles.

3.2 Weight Allocation and Alternative Evaluation

The key point of the evaluation is to be fast, easy to understand but also mathematically sound. Because of the characteristics of the mass-participation stakeholder group, stakeholders are often hard to reach, and they do not take the time to understand the methodology of the calculation, but focus on expressing their preference and priority. Also, non-technical stakeholders are difficult to understand the mathematical meaning of the evaluation methods [23]. Thus, PROSE is chosen. This method applies a weighted sum approach based on order statistics to combine the individual profile distribution. It is well suitable for mass-participation evaluation, as it does not considers only the mean distribution values, but also standard deviations [20].

Weight Allocation. An efficient and transparent weight elicitation technique proposed by Kunsch and Brans is applied in this model, which is based on semantic relative-importance classes; stakeholders are required to weigh the criteria based on their priorities [24]. They need to represent relative importance's on an ordinal score level: 1 (Least important), 2 (Less important), 3 (Middle), 4 (More Important), 5 (Most important). The scale is chosen based on the magic number 7 plus or minus 2; by choosing the 5-point Likert scale (LS), the stakeholders can have space of the mind to process the information [25]. In the meantime, the priority ranking has enough levels concerning the accuracy of the weighing. Plus, the "0" class (Not relative) is added for giving a vanishing weight in the judgment. Stakeholders are asked to define relative-importance classes in the above-mentioned scale. They need to rank at least one criterion as the "most important" as it is never empty. Then, stakeholders weigh the other criteria by comparing the most important criterion.

Weight allocations from all stakeholders in the group are collected. Suppose there are n criteria in the criteria set of the stakeholder group, the multiple-stakeholder profiles of criterion k rank on the class weight score i_c is w_{ki_c}, which means the proportionality of the criterion percentage profile of the class weights. By taking the arithmetic mean of the importance's classes, the not-normalized weight (NNW) of the criterion k is gotten:

$$NNW_k = \sum_{i_c=0}^{5} i_c \times w_{ki_c}; i_c = 0, 1, 2, 3, 4, 5 \tag{1}$$

Then the normalized weight (NW) of the criterion k is the NNW of criterion k proportional to the NNW set:

$$NW_k = \frac{NNW_k}{\sum_{j=1}^{n} NNW_j} \tag{2}$$

In this way, the global weight allocation of the stakeholders from the stakeholder group is calculated.

Alternative Evaluation. Suppose stakeholders have to evaluate a finite set of alternative $A = \{a_1, a_2, ..., a_m\}$, stakeholders are asked to give performance scores on the alternatives based on each criterion. A 5-point LS is used, and at least one alternative needs to be scored 5 as the "most preferred" for one criterion. The other alternatives are scored by comparing the most preferred alternative, which is treated as a benchmark. After collecting all the evaluation data, the performance percentage profile p_{tji} of alternative t on the class weight score i_a based on criterion j is gotten.

The calculation of the performance scores considers the profile distributions. to get the global performance indicator of an alternative a_t, say S_t, the global weight profile set $G_t = \{g_0, g_1, g_2, g_3, g_4, g_5\}$ needs to be calculated first:

$$G_t = \{g_{ti_a} = \sum_{j=1}^{n} NW_j \times p_{tji_a}\}; i_a = 0, 1, 2, 3, 4, 5 \tag{3}$$

Where i_a is the alternative performance score class. After obtaining the global weight profile set of one alternative, its global mean score V_t can be calculated:

$$V_t = \sum_{i_a=0}^{5} i_a \times g_{ti_a} \tag{4}$$

Still, the sole global mean score loses the important information concerning the profile dispersion, as the high deviation on the alternative performance scores will result in a nonconsensual solution among stakeholders. To obtain a safer ranking, the standard deviation of the performance score is considered. The standard deviation σ_t of V_t is given in:

$$\sigma_t = \sqrt{\sum_{i_a=0}^{5} g_{ti_a} \times (i_a - V_t)^2}$$ (5)

The final global performance indicator combines mean value and spread measured by the standard deviation:

$$S_t = V_t - \sigma_t$$ (6)

Only the lower value from the interval of the standard deviation σ_t is kept for being on the safe performance side.

The evaluation process of the MAMCA survey model is finished by now. The final weight allocation of the mass-participation stakeholder group can be used in the normal MAMCA evaluation process. However, it is advised not to include the alternative performance indicators as the final evaluation scores of the stakeholder group. Instead, the global performance indicators of alternatives should be treated as a reference to the stakeholders' preferences. It is believed that the criteria priority ranking is much more objective than the alternative evaluation. The alternative evaluation requires more objective data and information to support, so the process of the alternative evaluation needs to be executed preferably by the experts. Still, the decision-maker can compare the result of the evaluation of experts and the stakeholders' performance indicators for further investigating. E.g., they can have a discussion with the stakeholders on it to see what their potential misconception is, use it to determine communication focus on specific alternatives.

4   Case Study

In order to apply the MACMA survey model in practice, a survey tool is developed in the MACMA software. Dedicated pages for the survey tool are built, called "MAMCA survey tool" pages. Each MAMCA project has individual survey setting pages. And the decision-maker can publish the surveys dedicated to different stakeholder groups, in which different survey questions can be asked. Also, the decision-maker has an option to ask stakeholders to evaluate alternatives or not, while the weight allocation of criteria is a must.

To demonstrate the MAMCA mass-participation function, a fictive case entitled "The last-mile in the supply chain" is used. The case aimed to gain insight into the extent to which different alternatives for the last mile of a supply chain for home deliveries contribute to the interests of the different stakeholder groups involved. In this case study, there is a stakeholder group "citizens", that is suitable for validating the mass-participation function. In this study, only the stakeholder group "citizens" is focused upon. The data shown here are for demonstration reason only and are not the result of

Table 1. Criteria of stakeholder group "citizens"

Criterion	Criterion description	Direction of preference
Road safety	The low risk that a person using the urban road network will be (fatally) injured	Maximization
Air quality	Low concentration of particulate matter, NOx and SO2 in the air	Maximization
Urban accessibility	Reduce freight transport, less congestion	Maximization
Attractive urban environment	Attractive and livable urban environment for its citizens	Maximization
Low noise nuisance	Reduce noise nuisance of road transportation	Maximization

an actual survey that was performed among citizens. The criteria of the "citizens" group and the corresponding descriptions and directions of preference are shown in Table 1.

Before distributing the survey, a relevant question about the stakeholders' SES is raised: "Is there a significant difference on the criteria priority ranking between car owners and non-owners?". The decision-maker can ask these types of questions through the survey (see Fig. 6). Then, a survey page dedicated to this stakeholder group can be generated. Stakeholders need to rank the priority of the criteria. The decision-maker can choose if stakeholders are also allowed to evaluate the alternatives.

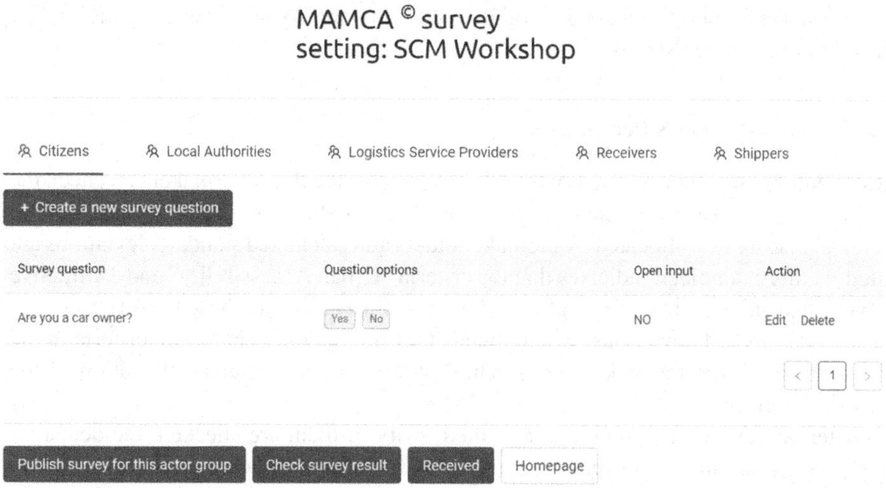

Fig. 6. The screenshot of MAMCA survey setting: design survey questions

4.1 Stakeholders' Perspective

The stakeholders receive the survey link that is sent by the decision-maker. The survey consists of 5 parts: Description of the project, overview of alternatives and criteria, answering survey questions (optional), weighing the criteria, evaluating the alternatives (optional). After going through the overview of the alternatives and criteria, they should answer the SES questions asked by the decision-maker. Next, the stakeholders need to give the importance scores to the criteria, and optionally, they will give the performance scores to the alternatives based on their preferences (see Fig. 7).

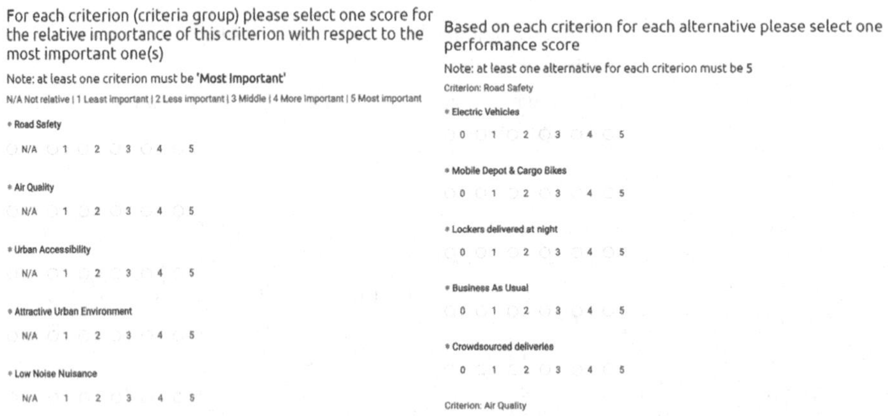

For each criterion (criteria group) please select one score for the relative importance of this criterion with respect to the most important one(s)

Note: at least one criterion must be 'Most Important'

N/A Not relative | 1 Least important | 2 Less important | 3 Middle | 4 More Important | 5 Most important

* Road Safety

 N/A   1   2   3   4   5

* Air Quality

 N/A   1   2   3   4   5

* Urban Accessibility

 N/A   1   2   3   4   5

* Attractive Urban Environment

 N/A   1   2   3   4   5

* Low Noise Nuisance

 N/A   1   2   3   4   5

Based on each criterion for each alternative please select one performance score

Note: at least one alternative for each criterion must be 5

Criterion: Road Safety

* Electric Vehicles

 0   1   2   3   4   5

* Mobile Depot & Cargo Bikes

 0   1   2   3   4   5

* Lockers delivered at night

 0   1   2   3   4   5

* Business As Usual

 0   1   2   3   4   5

* Crowdsourced deliveries

 0   1   2   3   4   5

Criterion: Air Quality

Fig. 7. Screenshots of the weight allocation and alternatives evaluation pages

The stakeholders do not need to log in to the software. By just answering the survey, the results will be registered.

4.2 Decision-Maker's Perspective

After invited stakeholders have finished the evaluation, the decision-maker can check the final result of the survey in the MAMCA software. As shown in Fig. 8, the table of the weights' distribution allocated by the stakeholders and calculated standard deviations are listed. In this example, it indicates that the criteria "Urban Accessibility" and "Attractive Urban Environment" have the highest NNWs; at the same time, these two criteria have the lowest standard deviations, which means they are the most important criteria in the points of view from the stakeholders. The NWs are the final weight allocation of the stakeholder group.

After all surveys are submitted and the quality of them are checked, the decision-maker can import the survey result to the MAMCA project by clicking one single button. The NWs of the survey will be treated as the weight allocation of the stakeholder group "citizens" and will be applied in the further evaluation of the MAMCA process.

Weights   Scores   Social-Economic profiles

Criteria Name ⇕	Not relative (0)	Least important (1)	Less important (2)	Middle (3)	More Important (4)	Most important (5)	Standard deviation	Not-normalized Weight	Normalised Weight
Road Safety	-	-	13.3%	53.3%	13.3%	20.0%	1.0	3.4	18.6%
Air Quality	-	6.7%	26.7%	13.3%	20.0%	33.3%	1.4	3.5	19.0%
Urban Accessibility	-	-	6.7%	13.3%	40.0%	40.0%	0.9	4.1	22.6%
Attractive Urban Environment	-	-	6.7%	26.7%	13.3%	53.3%	1.0	4.1	22.6%
Low Noise Nuisance	-	13.3%	20.0%	20.0%	33.3%	13.3%	1.3	3.1	17.2%

< 1 >

Copy survey weight data to project

Fig. 8. Screenshot of the "citizens" group's weight table

As mentioned before, in this case study we would like to investigate if the car owners in the group "citizens" would have a different rank of criteria priority than those who do not own a car. In the MAMCA survey tool, the decision-maker can add comparison groups based on asked survey questions (see Fig. 9). Two groups are created based on if the stakeholders own private cars. A pie chart showing the proportion of the answers indicates that the stakeholders who own private cars are slightly fewer than those who do not. A bar chart is generated that shows the weight allocation of the criteria from the two comparison groups. It can be seen there is a large difference in the importance of the criterion "Urban Accessibility", that the car owners rank as the most important criterion among all, while the other stakeholders rank it as the least important. Apart from that, the other importance of the criteria is similar. It makes sense that, the citizens overall find an attractive and livable urban environment important, but the car owners suffer from over-busy traffic so they also think less congestion is really important.

The decision-maker can have a further discussion on it, as now the "citizens" group has two different criteria priorities because of urban accessibility. Two sub-groups could be divided into the "citizens" group based on the SES "Private Car Ownership". The corresponding criteria weights are allocated regarding the SES. In the afterward MAMCA alternative evaluation, experts can give more rational evaluation scores for two sub-groups concern about their interests.

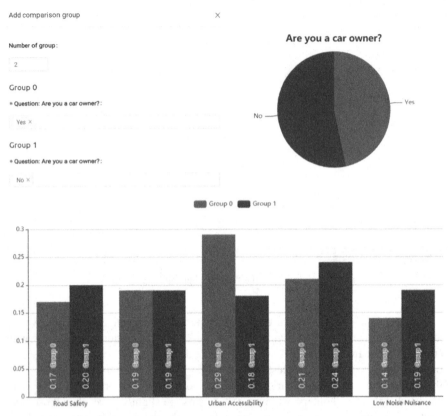

Fig. 9. Screenshots of the comparison function

5   Limitations and Directions for Future Research

This study tries to demonstrate the new MAMCA mass-participation survey tool. A fictive case is used in this study; it is a didactic case that was applied in the university. The students are the actors for different stakeholder groups as roleplays. In the end, 50 samples of the surveys are collected for the "citizens" stakeholder group. Still, there should be more responses of the voices as a mass-participation decision-making process. There are still a lot of potentials for this study. Two directions for the future research are listed below.

First is to have a study on a real "mass-participation" case. This paper mainly talks about the methodology of the mass-participation decision-making behind and focus on the presentation of the MAMCA survey tool. A mass-participation case in the real world concerns about sustainable urban construction logistics will be studied in the near future. By surveying the citizens in Brussels-Capital Region (BCR), Belgium, the opinions of the citizens can be gathered, and the mass-participation analysis can be applied. This mass-participation decision-making can be evaluated in this case.

Second is to have an in-depth discussion of the post process after gathering the survey. Due to a limited number of pages allowed, there is only a small discussion about the sub-group creation and evaluation after collecting the data. A dedicated work will be done to discuss when and how to regroup the stakeholder group or divided the group into sub-groups based on the collected information, e.g., standard deviations of the weight allocations, SES.

6 Conclusion

MAMCA methodology now shifts the concept of the "stakeholder" to the "stake-holder group", trying to hear the points of view from more stakeholders, instead of those of only one representative in each group. Elaborate types of groups like "citizens" have some characteristics that are inefficiently addressed by the current participation system. The stakeholders within this group are normally hard-to-reach and have quite different SES. To involve more stakeholders and hear the voices of them, a new MAMCA survey model for the mass-participation is designed. The survey model divides the tasks of the decision-maker and stakeholders, such that they can work singly instead of being gathered in the workshop. PROSE method is used for the evaluation process. It is a transparent method that applies a weighted sum approach based on order statistics to combine the individual profile distribution. It is suitable for the mass-participation evaluation as it is easy to understand but also mathematically sound. Additionally, the decision-maker can inquire about the SES of stakeholders for further investigation within the stakeholder group.

Following this, a survey tool built in MAMCA software is developed. The survey tool can explore more detail within one single stakeholder group. As there is a massive number of stakeholders participating, their priorities might be different. The survey tool not only indicates the weight allocation of the criteria, but also the standard deviation of the importance scores given. The decision-maker is able to find the homogeneity and heterogeneity within the stakeholder group: By creating comparison groups, the weight allocation of the criteria from stakeholders with different SES are displayed in a bar chart. If there is a significant difference in the ranking from the stakeholders with different SES, the decision-maker should consider regrouping or identifying sub-groups for the stakeholders.

References

1. Freeman, R.E.: Strategic Management: A Stakeholder Approach. Cambridge University Press, Cambridge (2010)
2. Banville, C., Landry, M., Martel, J.-M., Boulaire, C.: A stakeholder approach to MCDA. Syst. Res. Behav. Sci. Off. J. Int. Fed. Syst. Res. **15**, 15–32 (1998)
3. Browne, D., Ryan, L.: Comparative analysis of evaluation techniques for transport policies. Environ. Impact Assess. Rev. **31**, 226–233 (2011)
4. Macharis, C., de Witte, A., Ampe, J.: The multi-actor, multi-criteria analysis methodology (MAMCA) for the evaluation of transport projects: theory and practice. J. Adv. Transp. **43**(2), 183–202 (2009)
5. Lode, M.L.: Transition management and multi-actor multi-criteria analysis intertwined: how MAMCA supports transition management and its political potential. In: International Sustainability Transitions 2020 (2020)

6. Macharis, C., Verbeke, A., De Brucker, K.: The strategic evaluation of new technologies through multicriteria analysis: the ADVISORS case. Res. Transp. Econ. **8**, 443–462 (2004)
7. Macharis, C., Baudry, G.: Decision-Making for Sustainable Transport and Mobility: Multi Actor Multi Criteria Analysis. Edward Elgar Publishing, Cheltenham (2018)
8. Firouzabadi, S.A.K., Henson, B., Barnes, C.: A multiple stakeholders' approach to strategic selection decisions. Comput. Ind. Eng. **54**, 851–865 (2008)
9. Macharis, C., Turcksin, L., Lebeau, K.: Multi actor multi criteria analysis (MAMCA) as a tool to support sustainable decisions: state of use. Decis. Support Syst. **54**, 610–620 (2012)
10. Dodgson, J.S., Spackman, M., Pearman, A., Phillips, L.D.: Multi-criteria analysis: a manual. Department for Communities and Local Government: London (2009)
11. Macharis, C.: The importance of stakeholder analysis in freight transport. Eur. Transport **25–26**, 114–126 (2005)
12. Macharis, C., Kin, B., Lebeau, P.: Multi-actor multi-criteria analysis tool urban stakeholders to logistics involve. In: Urban Logistics: Management, Policy and Innovation in a Rapidly Changing Environment, p. 274 (2018)
13. Turcksin, L., et al.: A multi-actor multi-criteria framework to assess the stakeholder support for different biofuel options: the case of Belgium. Energy Policy **39**, 200–214 (2011)
14. Sun, H., Zhang, Y., Wang, Y., Li, L., Sheng, Y.: A social stakeholder support assessment of low-carbon transport policy based on multi-actor multi-criteria analysis: the case of Tianjin. Transp. Policy **41**, 103–116 (2015)
15. Keseru, I., Bulckaen, J., Macharis, C.: The use of AHP and PROMETHEE to evaluate sustainable urban mobility scenarios by active stakeholder participation: the case study of Leuven. In: 2nd International MCDA Workshop on PROMETHEE: Research and Case Studies (2015)
16. Huang, H., Lebeau, P., Macharis, C.: The multi-actor multi-criteria analysis (MAMCA): new software and new visualizations. In: Moreno-Jiménez, J.M., Linden, I., Dargam, F., Jayawickrama, U. (eds.) ICDSST 2020. LNBIP, vol. 384, pp. 43–56. Springer, Cham (2020). https://doi.org/10.1007/978-3-030-46224-6_4
17. Akter, T., Simonovic, S.P.: Aggregation of fuzzy views of a large number of stakeholders for multi-objective flood management decision-making. J. Environ. Manag. **77**, 133–143 (2005)
18. Irvin, R.A., Stansbury, J.: Citizen participation in decision making: is it worth the effort? Public Adm. Rev. **64**, 55–65 (2004)
19. Vehovar, V., Manfreda, K.L.: Overview: online surveys. In: The SAGE Handbook of Online Research Methods, vol. 1, pp. 177–194 (2008)
20. Kunsch, P.L., Ishizaka, A.: Multiple-criteria performance ranking based on profile distributions: an application to university research evaluations. Math. Comput. Simul. **154**, 48–64 (2018)
21. Socioeconomic status. American psychological association. https://www.apa.org/topics/socioeconomic-status. Accessed 30 Nov 2020
22. Liu, Y., Fan, Z.-P., You, T.-H., Zhang, W.-Y.: Large group decision-making (LGDM) with the participators from multiple subgroups of stakeholders: a method considering both the collective evaluation and the fairness of the alternative. Comput. Ind. Eng. **122**, 262–272 (2018)
23. Apostolakis, G.E., Pickett, S.E.: Deliberation: integrating analytical results into environmental decisions involving multiple stakeholders. Risk Anal. **18**, 621–634 (1998)
24. Kunsch, P.L., Brans, J.-P.: Visualising multi-criteria weight elicitation by multiple stakeholders in complex decision systems. Oper. Res. **19**(4), 955–971 (2019). https://doi.org/10.1007/s12351-018-00446-0
25. Miller, G.A.: The magical number seven, plus or minus two: some limits on our capacity for processing information. Psychol. Rev. **63**, 81 (1956)

26. Pita, C., Pierce, G.J., Theodossiou, I.: Stakeholders' participation in the fisheries management decision-making process: fishers' perceptions of participation. Mar. Policy **34**, 1093–1102 (2010)
27. Lourenco, R.P., Costa, J.P.: Incorporating citizens' views in local policy decision making processes. Decis. Support Syst. **43**, 1499–1511 (2007)

Using FITradeoff Method for Supply Selection with Decomposition and Holistic Evaluations for Preference Modelling

Lucia Reis Peixoto Roselli$^{(\boxtimes)}$ (iD) and Adiel Teixeira de Almeida (iD)

Center for Decision Systems and Information Development (CDSID), Universidade Federal de Pernambuco, Recife, PE, Brazil
{lrpr,almeida}@cdsid.org.br

Abstract. The FITradeoff method is a Flexible and Interactive method used to solve Multi-Criteria Decision Making/Aiding (MCDM/A) problems, with additive aggregation in the context of Multi-Attribute Value Theory. This study discusses the combination of two perspectives of preference modelling in the FITradeoff method for a supply selection decision problem. Five criteria are considered: Price, Product Quality, Delivery Time for supplying, Confidence of the Supplier and Service, associating to the classical objectives of manufacturing and operations strategies. The two perspectives are: the elicitation process by decomposition and the holistic evaluation. The combination of these two perspectives offers flexibility for the decision-maker during the FITradeoff decision process. The FITradeoff is implemented in a Decision Support System, in which the holistic evaluation is performed using graphical and tabular visualizations.

Keywords: FITradeoff method · Elicitation process · Holistic evaluation · Multi-Criteria Decision Making/Aiding (MCDM/A) · Multi-attribute · Supply selection

1 Introduction

The FITradeoff method [1, 2] is one of the methods that can be used to deal with Multi-Criteria Decision Making/Aiding (MCDM/A) problems in the context of MAVT (Multi-Attribute Value Thinking) [3–5]. Using one of those methods, including, the FITradeoff method, a solution can be obtained for MCDM/A problems, such as: the best alternative for choice problematic [1], the ranking of alternatives for ranking problematic [6], the classification of alternatives into categories for sorting problematic [7], and the best portfolio of alternatives for portfolio problematic [8]. The FITradeoff method is implemented in a Decision Support System (DSS) which is available by request at www.fitradeoff.org.

In literature, a wide range of applications have been performed using the FITradeoff to support several problems, named: energy selection [8–11], supplier selection problems [12, 13]; equipment selection [14], location problems [15, 16]; triage decision

© Springer Nature Switzerland AG 2021
U. Jayawickrama et al. (Eds.): ICDSST 2021, LNBIP 414, pp. 18–29, 2021.
https://doi.org/10.1007/978-3-030-73976-8_2

problem [17], scheduling decisions [18], in combination of the World-Class Manufacturing (WCM), and the Business Process Management (BPM) [19, 20], in water supply context [21], and in information system context [22].

Usually, multi-attribute methods uses either holistic or decomposition evaluation for preference modeling. However, using the FITradeoff method, the Decision-Maker (DM) can either use the elicitation by decomposition or the holistic evaluation in different steps of the decision process. In other words, in this method the combination of these two perspectives of preference modelling is available for DMs use, considering their cognitive style. This feature is originally discussed in the recent study of de Almeida et al. (2021) [2]. It is worth mentioning that although using holistic evaluation, this method has no relation with "preference disaggregation" approaches.

In this context, this study illustrates a decision-making process using the FITradeoff method to support a supplier selection problem. This problem presents five criteria, and it is associate to manufacturing and operations strategies. The contribution of the study is to shows how these two paradigms of decomposition or holistic evaluation can be combined in the FITradeoff method to solve a supplier selection problem, which is integrated with the manufacturing strategy approach.

This paper is organized as follows. Section 2 describes the FITradeoff method, Sect. 3 describes the supply selection problem, Sect. 4 discuss the FITradeoff process, and Sect. 5 draws some conclusions and suggestions for future studies.

2 Combining Holistic and Decomposition Evaluation in FITradeoff

The FITradeoff method [1, 2] combines two paradigms for preference modelling: the elicitation by decomposition and the holistic evaluation [2]. The elicitation by decomposition is based on the Tradeoff procedure [3] to elicit scaling constants in the context of MAVT [3]. The FITradeoff uses concepts of partial information, thus indifference relations between the consequences are not required to be established by the DM. According to Weber & Borcherding [23], the requirement to define indifference relations leads to 67% of inconsistencies in the results, obtained with the classical Tradeoff procedure. Also, using the FITradeoff method, a space of scaling constants is obtained, instead of the exact value of each scaling constant.

The FITradeoff method is considered as a Flexible and an Interactive method. The FITradeoff method offers flexibility for the DM to conduct the decision process.

In the elicitation process, initially the DM had to rank the scaling constants. After that, some elicitation questions are presented in order to compare pairs of consequences. The comparison of consequences followed the order of scaling constants, i.e., those consequences of adjacent criteria are compared. The elicitation questions are in the format of the elicitation question illustrated in Fig. 2.

In the holistic evaluation, the alternatives can be compared in a holistic way. The holistic evaluation is performed using graphical and tabular visualizations. In the FITradeoff Decision Support System (DSS), four types of visualizations are presented, named: bar graph, spider graph, bubble graph and table. Figures 3 and 4 illustrate bar graphs used to perform holistic evaluations.

In this context, considering these two perspectives, the DM can express preferences using the perspective that she/he judges as the most appropriated to her/his cognitive style. In other words, the DM can express preferences concerning the comparison of consequences or the comparison of alternatives, in each cycle of the process.

After each preference expressed by the DM, an inequality is generated, and it is included in a Linear Programing Problem (LLP). Thus, after each interaction with the DM, the LPP model runs, and the space of scaling constants can be reduced. Therefore, some dominance relations can be established, and partial results can be obtained during the FITradeoff decision process.

The FITradeoff method is considered interactive, since the DM participates on the whole process expressing preferences. Moreover, considering the elicitation process and the holistic evaluation, the FITradeoff process may be shorten since more inequalities can be included in the LPP model.

Therefore, in the FITradeoff method the two perspectives of preference modeling can be combined in an integrated way, providing flexibility for the DM. The DM can alternate between them (elicitation process and the holistic evaluation) during the decision process, performing those that she/he judges more appropriated to express preferences. The elicitation process by decomposition is performed by the comparison of pairs of consequences. On the other hand, in the holistic evaluation, alternatives are compared in a holistic way.

For instance, the DM can start the process performing the elicitation process by decomposition, i.e., answering some elicitation questions in the format of Fig. 2. Also, after each updating on the partial results, the DM can use the visualizations to evaluate the alternatives. Thus, based on the disposition of the alternatives, the DM can decide if a preference relation can be established between the alternatives, based on the holistic evaluation. Thus, for each inequality generated, the LPP model runs and the partial results can be updated, until a final solution has been obtained or the DM decides to interrupt the process [2]. It is worth mentioning that behavioral studies have been conducted to investigate how DMs perform the elicitation process by decomposition and the holistic evaluation [24–32].

In Section four (4), a decision process conducted with the FITradeoff method is described in order to illustrate the combination of these two paradigms of preference modelling. The FITradeoff method is available by request at www.fitradeoff.org.

3 Supply Selection Problem

There are many reported studies in the literature showing the use of MCDM/A for Supply Selection problems [12, 13, 33–35]. This particular selection problem is a case subsequently applied in order to illustrates the use of the FITradeoff method.

The problem includes seventeen (17) alternatives and five criteria: Price, Product Quality, Delivery Time for supplying, Confidence of the Supplier and Service. The first four criteria are associated to the classical objectives studied in manufacturing (and operation) strategy [36–38]. The criteria Service is associated to the capacity and to the quality of service that the supplier is able to supply for pos-delivering the product. This criterion is evaluated by experts using a Likert scale of five levels, in which the level five represents the highest performance and the level one represents the worst performance.

The Price is given in a range of US $50,000 to US $80,000 and it is naturally a decreasing in preference criteria. The Product Quality criterion is evaluated by experts using a Likert scale of five levels.

The Delivery Time and the Confidence are associated to strategic objectives, as it is considered in manufacturing and operation strategy studies [36–38]. The former is associated to the production and logistic capacity of the supplier and it is found in a range of 7 to 20 days. The latter is associated to the credibility of a supplier in delivering the product on time. That is, it is related to the ability a supplier has of accomplishing the committed Delivery Time. This criterion is evaluated by experts (range of 0 to 100) and represents the probability that the supplier can deliver the product on time. Table 1 shows the consequence matrix, with the consequences for each criterion for the 17 suppliers.

This case study is based on previous studies considering manufacturing strategy approach. Thus, the data presented in Table 1 are based on previous studies, although the data are prepared only for illustrating purpose in this case.

Table 1. Supplier selection decision matrix

Alternative vs Criteria	Confidence	Quality	Price	Delivery time	Service
Supplier 1	84	2	69057	14	2
Supplier 2	80	2	78954	10	5
Supplier 3	76	2	60625	14	5
Supplier 4	89	4	63572	11	2
Supplier 5	77	2	56036	18	4
Supplier 6	85	2	71143	20	1
Supplier 7	92	3	69571	7	2
Supplier 8	93	2	63215	18	3
Supplier 9	74	5	54414	13	1
Supplier 10	80	5	57043	18	3
Supplier 11	71	5	57288	12	3
Supplier 12	88	2	50633	19	3
Supplier 13	78	3	53063	9	3
Supplier 14	76	2	52848	19	2
Supplier 15	82	4	66776	19	2
Supplier 16	90	2	79240	13	4
Supplier 17	92	4	66591	20	5

4   Analyzing the Supply Selection Problem with FITradeoff DSS

The ranking of the suppliers has been obtained using the FITradeoff method for ranking problematic [2, 6]. The ranking is obtained based on dominance relations between the alternatives, computed from Eq. (1). Equation (1) is the objective function in the LPP model. The constrains are the preferences expressed by the DM during the decision process.

$$Max\,D(SupN, SupZ) = \sum_{i=1}^{n} k_i v_i (SupN_i) - \sum_{i=1}^{n} k_i v_i (SupZ_i) \tag{1}$$

The first preference relation expressed by the DM during the decision process is the ranking of the scaling constants, as illustrated in the Eq. (2).

$$K_{Confidence} > K_{Quality} > K_{Price} > K_{DeliveryTime} > K_{Service} \tag{2}$$

Thus, this inequality is included in the LPP model and after that, some dominance relations have been observed between the suppliers. The suppliers 4, 7, 8, 10, and 17 are incomparable in the top of the ranking and dominated the other. However, all the suppliers continue allocated in the same position of the ranking, as illustrated by the Hasse Diagram in Fig. 1. The Hasse Diagram is obtained in the FITradeoff DSS.

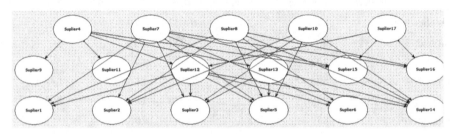

Fig. 1. Hasse Diagram with one position

In this context, in order to obtain more information about the suppliers' ranking, the DM continue the elicitation process by decomposition, comparing the consequences in pairs. The first elicitation question answered by the DM is illustrated in Fig. 2.

In this case, the DM prefers a supplier which had intermediate performance in the criterion Confidence, then a suppler which had the best performance in the criterion Service. Thus, the inequality in the format of Eq. (3) is inserted in the LPP model.

$$K_{Confidence} v_i(x_i)) > K_{Service} \tag{3}$$

After seven elicitation questions, in the same format of Fig. 2, the ranking has been updated. The Supplier 6 has been defined as the worst of the group, i.e. the supplier 6 has been dominated by the other, in a transitive way. On the other hand, the Suppliers 4, 10 and 17 are incomparable in the top of the ranking.

The DM can consider the holistic evaluation in order to compare the suppliers that are incomparable in the positions of the ranking. However, the DM decide to continue the elicitation process by decomposition since the number of alternatives which are incomparable are high, and several holistic evaluations should be conducted.

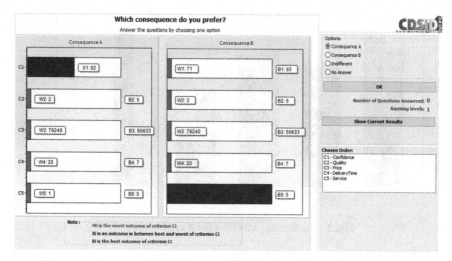

Fig. 2. Elicitation question

Therefore, after more four elicitation questions, the ranking was updated to six positions. At this moment, the Suppliers 4 and 17 are incomparable in the first position, and the Suppliers 7 and 10 are incomparable in the second position. Thus, the DM wishes to perform the holistic evaluation to compare these suppliers. The FITradeoff DSS offers four types of visualizations to conduct the holistic evaluation, named: bar graph, spider graph, bubble graph and table. The DM prefers the bar graphic since it is quite common in decision-making process. Also, based on neuroscience studies the bar graphs have been indicated as the most appropriated kind of visualization [28, 31, 32]. Thus, Fig. 3 illustrates the bar graph used to compare the Suppliers 4 and 17, and Fig. 4 illustrates the bar graph used to compare the Suppliers 7 and 10. In these bar graphs, the heights of the bars represent the performance of the alternatives.

Based on Fig. 3, the DM observes that the performances of Suppliers 4 and 17 are remarkably similar in the first three criteria (Confidence, Quality and Price). Hence, the DM does not feel comfortable to define a domination between them, based on the holistic evaluation. On the other hand, based on Fig. 4, the DM consider the Supplier 10 preferable than the Suppler 7, since the former presents the highest performances in the criteria Quality and Price. Also, the Supplier 10 presents similar performance in the first criterion (Confidence), nearly to 80% of the performance of the Supplier 7. Therefore, this preference relation is included in the LPP model, in the format of Eq. (4), and the Hasse Diagram has been updated, as illustrated in Fig. 5.

$$\sum_{i=1}^{n} k_i v_i(Sup10_i) > \sum_{i=1}^{n} k_i v_i(Sup7_i) \tag{4}$$

In order to obtain more information about the first position of the ranking, the DM decides to continue the elicitation process by decomposition. Thus, more five elicitation questions have been answered.

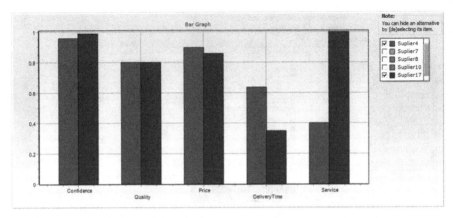

Fig. 3. Bar Graphic with Supplier 4 and Supplier 17

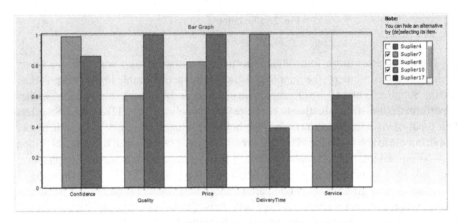

Fig. 4. Bar Graphic with Supplier 7 and Supplier 10

After that, the ranking has been updated, presenting thirteen (13) positions. In this new version of the ranking, the Supplier 4 is the one in the first position. Moreover, the Suppliers 9 and 11 are incomparable in the third position, and the Suppliers 1, 2 and 5 are incomparable in the eleventh position.

In this context, the DM wishes to compare, in a holistic way, the Suppliers 9 and 11, because they are in the top of ranking. Figures 6 illustrates the bar graph with the Suppliers 9 and 11. On the other hand, the DM do not wish to compare the suppliers 1, 2 and 5 because they are in the final of the ranking.

Based on Fig. 6, the DM observes that the suppliers present same performance on the criterion Quality. Also, the suppliers present similar performance on the criteria Confidence and Price, but the supplier 9 presents a small advantage in performance compared to the supplier 11. On the other hand, the supplier 11 presents a small advantage in performance in the criterion Delivery Time, and a marked advantage in criterion

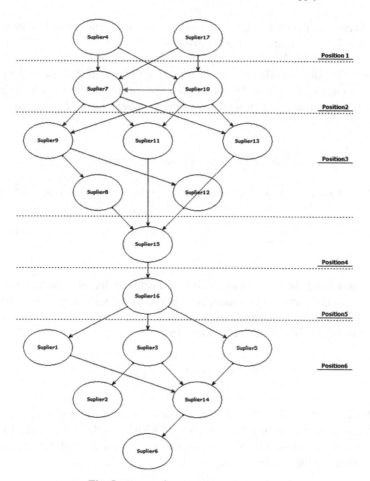

Fig. 5. Hasse Diagram with six positions

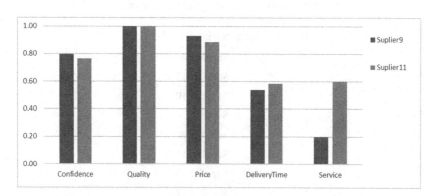

Fig. 6. Bar Graphic with Supplier 9 and Supplier 11

Service. Therefore, the DM judges this visualization as a quite difficult to establish a dominance relation between these alternatives.

In this context, the DM wishes to consider other visualizations presented in the FITradeoff DSS. The DM selected the table to use. In FITradeoff DSS, the tables present the decision matrix data (Table 1), but only for the specific alternatives in evaluation. Thus, Table 2 illustrates the performances of the Suppliers 9 and 11 in each one of the criteria.

Table 2. Table with the Suppliers 9 and 11

Alternative vs Criteria	Confidence	Quality	Price	Delivery time	Service
Supplier 9	74	5	54414	13	1
Supplier 11	71	5	57288	12	3

Based on Table 2, the DM considered that the small advantages of the Supplier 9 over the Supplier 11 in the criteria Confidence and Price, are sufficient to prefer the Supplier 9. The supplier 9 is those that presents the small price in the set. Thus, the inequality in the format of Eq. (5) is included in the LPP model.

$$\sum_{i=1}^{n} k_i v_i (Sup9_i) > \sum_{i=1}^{n} k_i v_i (Sup11_i) \tag{5}$$

At this point, the DM stops the decision process since the first ten position had been defined, as illustrated in Table 3. If the DM wishes, she/he can continue the decision process in the FITradeoff DSS, considering the combination of the elicitation by decomposition and the holistic evaluation.

Table 3. First ten positions of the ranking

Position	Supplier
1	Supplier 4
2	Supplier 17
3	Supplier 10
4	Supplier 7
5	Supplier 9
6	Supplier 11
7	Supplier 13
8	Supplier 12
9	Supplier 8
10	Supplier 15

It is worth mentioning that the sequence of preferences expressed during the decision process, in the elicitation or the holistic evaluation, depends on the DM. If other DM performs the same problem, and present different preferences for the comparisons, the partial order should be different.

5 Conclusion

Usually, multi-attribute methods uses either holistic or decomposition evaluation for preference modeling. In each case it is assumed that one of them is more appropriate for the decision process. However, in the FITradeoff method, the DM can choose the most appropriate perspectives - elicitation process by decomposition and the holistic evaluation, in each step of the decision process. For instance, in the elicitation process by decomposition the DM express preferences for pairs of consequences. On the other hand, in the holistic evaluation the DM express preferences for pairs of alternatives [2].

Therefore, this study illustrates how the new features of the FITradeoff method works in the supply selection problem, which is a relevant kind of decision problem, often approached in the literature. Previous studies with FITradeoff method considered the supply selection problem [12, 13], but in this study the problem is integrated with the manufacturing strategy approach.

Several behavioral studies have been performed and are already publish in the literature concerning to the FITradeoff method. These studies investigate behavioral aspects in the elicitation process and the holistic evaluation. Thus, for future research, additional behavioral studies should be performed in order to investigate the combination of these paradigms in the FITradeoff decision process, since the previous one investigated the elicitation process and the holistic evaluation in an individual way.

Acknowledgment. This work had partial support from the Brazilian Research Council (CNPq).and FACEPE (Foundation for Research in the State of Pernambuco). Grant Nos. APQ-0484-3.08/17, APQ-0370-3.08/14.

References

1. de Almeida, A.T., Almeida, J.A., Costa, A.P.C.S., Almeida-Filho, A.T.: A new method for elicitation of criteria weights in additive models: flexible and interactive tradeoff. Eur. J. Oper. Res. **250**(1), 179–191 (2016)
2. de Almeida, A.T., Frej, E.A., Roselli, L.R.P.: Combining holistic and decomposition paradigms in preference modeling with the flexibility of FITradeoff. CEJOR **29**(1), 7–47 (2021). https://doi.org/10.1007/s10100-020-00728-z
3. Keeney, R.L., Raiffa, H.: Decisions with Multiple Objectives: Preferences, and Value Tradeoffs. Wiley, New York (1976)
4. Belton, V., Stewart, T.: Multiple Criteria Decision Analysis. Kluwer Academic Publishers, Dordrecht (2002)
5. Figueira, J., Greco, S., Ehrogott, M.: Multiple Criteria Decision Analysis: State of the Art Surveys. Springer, New York (2005). https://doi.org/10.1007/b100605
6. Frej, E.A., de Almeida, A.T., Costa, A.P.C.S.: Using data visualization for ranking alternatives with partial information and interactive tradeoff elicitation. Oper. Res. Int. J. **19**(4), 909–931 (2019). https://doi.org/10.1007/s12351-018-00444-2

7. Kang, T.H.A., Frej, E.A., de Almeida, A.T.: Flexible and interactive tradeoff elicitation for multicriteria sorting problems. Asia Pac. J. Oper. Res. **37**, 2050020 (2020)
8. Frej, E.A., Ekel, P., de Almeida, A.T.: A benefit-to-cost ratio based approach for portfolio selection under multiple criteria with incomplete preference information. Inf. Sci. **545**, 487–498 (2021)
9. Fossile, D.K., Frej, E.A., da Costa, S.E.G., Pinheiro, E., de Lima, A., de Almeida, T.: Selecting the most viable renewable energy source for brazilian ports using the FITradeoff method. J. Clean. Prod. **260**, 121107 (2020). https://doi.org/10.1016/j.jclepro.2020.121107
10. Kang, T.H.A., Júnior, A.M.D.C.S., de Almeida, A.T.: Evaluating electric power generation technologies: a multicriteria analysis based on the FITradeoff method. Energy **165**, 10–20 (2018)
11. de Macedo, P.P., de Miranda Mota, C.M., Sola, A.V.H.: Meeting the Brazilian energy efficiency law: a flexible and interactive multicriteria proposal to replace non-efficient motors. Sustain. Urban Areas **41**, 822–832 (2018)
12. Santos, I.M., Roselli, L.R.P., da Silva, A.L.G., Alencar, L.H.: A supplier selection model for a wholesaler and retailer company based on FITradeoff multicriteria method. Math. Probl. Eng. **2020**, 1–14 (2020). https://doi.org/10.1155/2020/8796282
13. Frej, E.A., Roselli, L.R.P., Araújo de Almeida, J., de Almeida, A.T.: A multicriteria decision model for supplier selection in a food industry based on FITradeoff method. Math. Probl. Eng. **2017**, 1–9 (2017). https://doi.org/10.1155/2017/4541914
14. Carrillo, P.A.A., Roselli, L.R.P., Frej, E.A., de Almeida, A.T.: Selecting an agricultural technology package based on the flexible and interactive tradeoff method. Ann. Oper. Res. 1–16 (2018)
15. Dell'Ovo, M., Oppio, A., Capolongo, S.: Decision Support System for the Location of Healthcare Facilities: SitHealth Evaluation Tool. PoliMI SpringerBriefs, 1st edn, vol. XII, pp. 121. Springer, Cham (2020). https://doi.org/10.1007/978-3-030-50173-0
16. Camara e Silva, L., Daher, S.D.F.D., Santiago, K.T.M., Costa, A.P.C.S.: Selection of an integrated security area for locating a state military police station based on MCDM/A method. In: 2019 IEEE International Conference on Systems, Man and Cybernetics (SMC), pp. 1530–1534, October 2019
17. Camilo, D.G.G., de Souza, R.P., Frazão, T.D.C., da Costa Junior, J.F.: Multi-criteria analysis in the health area: selection of the most appropriate triage system for the emergency care units in natal. BMC Med. Inform. Decis. Mak. **20**(1), 1–16 (2020)
18. Pergher, I., Frej, E.A., Roselli, L.R.P., de Almeida, A.T.: Integrating simulation and FITradeoff method for scheduling rules selection in job-shop production systems. Int. J. Prod. Econ. **227**, 107669 (2020)
19. Silva, M.M., de Gusmão, A.P.H., de Andrade, C.T.A., Silva, W.: The integration of VFT and FITradeoff multicriteria method for the selection of WCM projects. In: 2019 IEEE International Conference on Systems, Man and Cybernetics (SMC), pp. 1513–1517. IEEE (2019)
20. Lima, E.S., Viegas, R.A., Costa, A.P.C.S.: A multicriteria method based approach to the BPMM selection problem. In: 2017 IEEE International Conference on Systems, Man, and Cybernetics (SMC), pp. 3334–3339. IEEE (2017)
21. Monte, M.B.S., Morais, D.C.: A decision model for identifying and solving problems in an urban water supply system. Water Resour. Manag. **33**(14), 4835–4848 (2019). https://doi.org/10.1007/s11269-019-02401-w
22. e Gusmao, A.P.H., Pereira Medeiros, C.: A model for selecting a strategic information system using the FITradeoff. Math. Probl. Eng. **2016**, 1–7 (2016). https://doi.org/10.1155/2016/7850960
23. Weber, M., Borcherding, K.: Behavioral influences on weight judgments in multi-attribute decision making. Eur. J. Oper. Res. **67**, 1–2 (1993)

24. Silva A.L.C.L, Costa, A.P.C.S., de Almeida, A.T.: Exploring cognitive aspects of FITradeoff method using neuroscience tools. Ann. Oper. Res. 1–23 (2021). https://doi.org/10.1007/s10 479-020-03894-0

25. de Almeida, A.T., Roselli, L.R.P.: NeuroIS to improve the FITradeoff decision-making process and decision support system. In: Davis, F.D., Riedl, R., vom Brocke, J., Léger, P.-M., Randolph, A.B., Fischer, T. (eds.) NeuroIS 2020. LNISO, vol. 43, pp. 111–120. Springer, Cham (2020). https://doi.org/10.1007/978-3-030-60073-0_13

26. Reis Peixoto Roselli, L., de Almeida, A.T.: Analysis of graphical visualizations for multicriteria decision making in FITradeoff method using a decision neuroscience experiment. In: Moreno-Jiménez, J.M., Linden, I., Dargam, F., Jayawickrama, U. (eds.) ICDSST 2020. LNBIP, vol. 384, pp. 30–42. Springer, Cham (2020). https://doi.org/10.1007/978-3-030-462 24-6_3

27. Roselli, L.R.P., de Almeida, A.T.: Improvements in the FITradeoff decision support system for ranking order problematic based in a behavioral study with NeuroIS tools. In: Davis, F.D., Riedl, R., vom Brocke, J., Léger, P.-M., Randolph, A.B., Fischer, T. (eds.) NeuroIS 2020. LNISO, vol. 43, pp. 121–132. Springer, Cham (2020). https://doi.org/10.1007/978-3-030-60073-0_14

28. Roselli, L.R.P., de Almeida, A.T., Frej, E.A.: Decision neuroscience for improving data visualization of decision support in the FITradeoff method. Oper. Res. Int. J. 19(4), 933–953 (2019). https://doi.org/10.1007/s12351-018-00445-1

29. Roselli, L.R.P., Pereira, L., da Silva, A., de Almeida, A.T., Morais, D.C., Costa, A.P.C.S.: Neuroscience experiment applied to investigate decision-maker behavior in the tradeoff elicitation procedure. Ann. Oper. Res. 289(1), 67–84 (2019). https://doi.org/10.1007/s10479-019-03394-w

30. Carneiro de Lima da Silva, A.L., Cabral Seixas Costa, A.P.: FITradeoff decision support system: an exploratory study with neuroscience tools. In: Davis, F.D., Riedl, R., vom Brocke, J., Léger, P.-M., Randolph, A., Fischer, T. (eds.) Information Systems and Neuroscience. LNISO, vol. 32, pp. 365–372. Springer, Cham (2020). https://doi.org/10.1007/978-3-030-28144-1_40

31. Roselli, L.R.P., Frej, E.A., de Almeida, A.T.: Neuroscience experiment for graphical visualization in the FITradeoff decision support system. In: Chen, Ye., Kersten, G., Vetschera, R., Xu, H. (eds.) GDN 2018. LNBIP, vol. 315, pp. 56–69. Springer, Cham (2018). https://doi.org/10.1007/978-3-319-92874-6_5

32. de Almeida, A.T., Roselli, L.R.P.: Visualization for decision support in FITradeoff method: exploring its evaluation with cognitive neuroscience. In: Linden, I., Liu, S., Colot, C. (eds.) ICDSST 2017. LNBIP, vol. 282, pp. 61–73. Springer, Cham (2017). https://doi.org/10.1007/978-3-319-57487-5_5

33. Chai, J., Liu, J., Ngai, E.: Application of decision-making techniques in supplier selection: a systematic review of literature. Expert Syst. Appl. 40(10), 3872–3885 (2013)

34. Xia, W., Wu, Z.: Supplier selection with multiple criteria in volume discount environments. Omega 35, 494–504 (2007)

35. Barla, S.B.: A case study of supplier selection for lean supply by using a mathematical model. Logist. Inf. Manag. 16, 451–459 (2003)

36. Slack, N., Lewis, M.: Operations Strategy. Prentice Hall, Upper Saddle River (2002)

37. Hill, T.: Operations Management – Strategy Context and Managerial Analysis, 2nd edn. Macmillan, New York (2000)

38. Hill, T.: Manufacturing Strategy, 2nd edn. Macmillan, New York (1993)

DEX2Web – A Web-Based Software Implementing the Multiple-Criteria Decision-Making Method DEX

Adem Kikaj[1,2(✉)] and Marko Bohanec[2(✉)]

[1] Jožef Stefan International Postgraduate School, Jamova 39, 1000 Ljubljana, Slovenia
adem.kikaj@ijs.si
[2] Department of Knowledge Technologies, Jožef Stefan Institute, Jamova 39, 1000 Ljubljana, Slovenia
marko.bohanec@ijs.si

Abstract. DEX2Web is an online suite of tools to help individuals and groups with their decision-making. DEX2Web implements the qualitative multiple-criteria decision-modelling method DEX. DEX is useful for supporting complex decision-making tasks, where there is a need to select a particular option from a set of possible ones to satisfy the goals of the decision-maker. DEX2Web primarily supports interactive development and evaluation of DEX models. Most of the functionality of the first available version of DEX2Web is inherited from its desktop ancestor DEXi: development of DEX model structure, editing of attributes and their scales, definition of decision rules, multi-attribute evaluation and analysis of alternatives, and presenting evaluation results with charts. DEX2Web has a modern software architecture and employs a newly developed DEX software library. DEX2Web is freely available on https://dex2web.ijs.si/.

Keywords: DEX2Web · DEX · Decision-making software · Multiple-criteria · Web · Software · Group decision-making

1 Introduction

Decision-making is the process of identifying and choosing alternatives based on the values, preferences, and beliefs of the decision-maker. Some decision problems are inherently difficult because of various obstacles, such as missing information, uncertainty, conflicting goals, and opposing views of multiple decision-makers. In such situations, decision-making may substantially benefit from using decision-aiding methods and tools.

Decision problems of a type that involve multiple, possibly conflicting criteria, can be aided using *Multiple-Criteria Decision-Making (MCDM)* methods [1, 2]. The aim is to help the decision maker understand better and structure a decision problem to represent it in the form of a decision model and use this model for decision-making tasks, such as choosing, ranking/sorting decision alternatives, and analyzing the gained results [2].

© Springer Nature Switzerland AG 2021
U. Jayawickrama et al. (Eds.): ICDSST 2021, LNBIP 414, pp. 30–43, 2021.
https://doi.org/10.1007/978-3-030-73976-8_3

To support the use of MCDM methods in practice, various multi-criteria *decision-making software (DMS)* have been developed [3, 4]. Decision-making software usually provides ways to build a decision model and analyze the results. Decision-making software consists of various forms of computer programs designed to enable users to process a set of goals to be achieved, alternatives available for making them, and relations between goals and options. *World Wide Web (WWW)* technologies have rapidly transformed the design, development, and implementation of all types of DMS [5, 6]. In this work, we are focused on the decision-making method DEX and its implementation as a web-based DMS.

Decision Expert (DEX) is a qualitative MCDM method. Currently, the DEX method is implemented in freely available software called *DEXi* [10]. DEXi is useful for supporting complex decision-making tasks, and it runs as a desktop application on the Microsoft Windows platforms. DEXi supports two primary functions: (1) development of qualitative multi-attribute models and (2) applying these models for the evaluation and analysis of decision options. DEX software is also distributed partly in other computing platforms. DEXi has been extensively and successfully used in numerous national and international projects [7].

DEXi software was released 20 years ago. Even though it has been regularly updated and extended with new features, it has reached the state that calls for a thorough renovation. In this work, we are designing modern web-based software called *DEX2Web* with an enhanced architecture that integrates the DEX method through the newly developed *DEX library*.

The structure of this paper is as follows. The next two sections describe the DEX method and supporting software, respectively. The implementation of DEX2Web is presented in Sect. 4. In Sect. 5, we conclude this work.

2   Qualitative Multiple-Criteria Method DEX

DEX [7–9, 11, 12] is a qualitative, hierarchical, multi-criteria decision-modelling method for the evaluation and analysis of decision alternatives. DEX decision models have a hierarchical structure, representing a decomposition of some decision problems into smaller, less complex sub-problems. DEX models are developed by defining (i) attributes, (ii) attribute scales, (iii) hierarchically structure of the attributes, and (iv) decision rules [13].

The DEX model thus consists of:

- *Attributes*: variables that represent basic features and assessed values of decision alternatives. DEX models can have one or more *root* attribute(s), i.e., those that do not have any *parent* attributes. Attributes that do not have any *child* attributes are called *input* attributes; all other attributes are *aggregated* attributes.
- *Scales* of attributes: these are *qualitative* scales and consist of a set of words, such as: "excellent", "acceptable", "inappropriate", etc. Usually, but not necessarily, scales are *ordered preferentially*, i.e., from bad to good values.
- *Hierarchy* of attributes: represents the decomposition of the decision problem and relations between attributes; higher-level attributes depend on lower-level ones. In general, a hierarchy is a directed graph without cycles.

- *Aggregation* function: to aggregate qualitative values, DEX primarily uses decision tables [13], which can be interpreted as collections of *if-then* rules. To define an aggregation function, the decision-maker defines an output value for each combination of input attribute's values.

Decision alternatives (also called *options* in DEXi) are defined by qualitative values, which are assigned to the model's input attributes. DEX evaluates alternatives in a bottom-up way, progressively aggregating the values according to the hierarchical structure of the model. Values of aggregated attributes, for which their respective children attributes already have values assigned, are aggregated using the corresponding aggregating functions. The final evaluation of an alternative is represented by the values calculated at the model's roots.

The DEX model that will be illustrated in this section will be used throughout this work to describe different properties of the DEX method. The model described here is the well-known DEX model *Car* [14], which is distributed together with the DEXi software.

Table 1. On the left side, the structure of the attributes is presented, and on the right side, their corresponding scales are displayed. The scales are ordered from worst values (shown in red) to best values (shown in green).

Attribute	Scale
CAR	unacc, **acc**, **good**, exc
├─PRICE	high, **medium**, low
│ ├─BUY.PRICE	high, **medium**, low
│ └─MAINT.PRICE	high, **medium**, low
└─TECH.CHAR.	bad, **acc**, **good**, exc
├─COMFORT	small, **medium**, high
│ ├─#PERS	to_2, **3-4**, more,
│ ├─#DOORS	2, 3, **4**, more
│ └─LUGGAGE	small, **medium**, big
└─SAFETY	small, **medium**, high

The Car model is a simple model for evaluating cars, where the input attributes influence only one parent attribute. All the attributes have qualitative values assigned, which is shown in Table 1. There are 10 attributes where 6 are *input*, and 4 are *aggregated*, including one *root* attribute.

Table 2 shows the *aggregation function* (decision rules) of the *CAR* attribute. The two leftmost columns correspond to the two attributes that influence *CAR*: *PRICE* and *TECH.CHAR.* These two columns contain all possible combinations of their input values. The third column gives the *output* value of the aggregation function for the respective row. The third column is filled-in by the decision-maker.

Hereafter, this paper is focused on the implementation of the DEX2Web software. For further methodological issues and recommendations for DEX model development, for instance, how to choose and structure attributes and how to define value scales, the reader is referred to [7–10].

Table 2. Decision table of the aggregated *CAR* attribute.

	PRICE	*TECH.CHAR.*	*CAR*
1	high	bad	**unacc**
2	high	acc	**unacc**
3	high	good	**unacc**
4	high	exc	**unacc**
5	medium	bad	**unacc**
6	medium	acc	**acc**
7	medium	good	**good**
8	medium	exc	**exc**
9	low	bad	**unacc**
10	low	acc	**good**
11	low	good	**exc**
12	low	exc	**exc**

3 Software for DEX

Three main generations of DEX qualitative modelling computer programs have been developed so far:

1. *DECMAK* [15] was released in 1981 for operating systems RT-11, VAX/VMS, and later for MS-DOS.
2. *DEX* [16] was released in 1987 as an integrated interactive computer program for MS-DOS.
3. *DEXi* [10] was released in 2000 as an interactive educational program for MS-Windows.

Over the years, additional features were added to DEXi, which gradually became a complete, stable and *de-facto* standard implementation of the DEX method. DEXi supports an interactive creation and editing of all components of DEX models (attributes, their hierarchy and scales, decision tables and alternatives), and provides methods for the evaluation and analysis of alternatives (what-if analysis, "plus-minus-1" analysis, selective explanation, comparison of alternatives). DEXi is free software, available at https://kt.ijs.si/MarkoBohanec/dexi.html. There are other implementations related to DEX, for instance, *proDEX* [17], a stand-alone Python implementation of DEX, and DEXx [9], a Java-based library.

DEX library is a new software library, developed from scratch, that implements all the DEX features proposed through decades of development in a unified, compact and modern software architecture. The library's main components have been designed according to the proposal of the so-called *Extended DEX* [9] and include:

- *DexProjects* is a new class aimed at managing multiple DEX models within the same decision project.
- *DexModels* represents a single DEX model.
- *DexAttributes* represents a single attribute (variable) of a DEX model. Attributes can be structured into trees or true hierarchies (directed acyclic graphs).
- *DexScales* represents attribute value scales. In addition to qualitative scales, which are exclusively used in DEXi, the new library also supports bounded and unbounded integer and continuous scales.
- *DexValues* is a class that represents different values that can be assigned to attributes. In addition to qualitative values used in DEXi, the DEX library supports numerical values, intervals, sets, probabilistic and fuzzy distributions, and offsets.
- *DexFunctions* represents aggregation functions. In comparison with DEXi, a number of new function types have been introduced, including constant, weighted, marginal, discretization and programmable functions.
- *DexAlternatives* is a component that represents decision alternatives.
- *DexEvaluation* provides the functionality needed to evaluate the existing alternatives of a DEX model, using various value propagation methods: single value, interval, set, or probabilistic or fuzzy value distribution.
- *DexViews* represents different states of objects of classes of the library that need to be represented in the user interface. The output of a *DexView* can be a JSON or XML object.
- *DexEditors* contain a collection of classes for editing components of DEX models: projects, models, scales, aggregation functions and alternatives.

The DEX library is used as the core component of DEX2Web.

4 DEX2Web

The main purpose of DEX2Web is to provide a new modern software supporting the DEX method. DEX2Web employs the DEX library and provides a web-based user interface so that the software can be used through an internet browser. In the current version, which is presented in this paper, we primarily implemented the functionality that is currently available in DEXi. The web-based environment also encouraged us to add some *group decision-making* features.

In this section, we describe the functional and non-functional requirements for the software, followed by a description of software architecture and an example of using the system.

4.1 Functional Requirements

The functional requirements of the DEX2Web are defined as follows:

- *Usage of the DEX2Web* – DEX2Web allows the users to create or import a single DEX project, which may contain multiple DEX models. DEX projects are stored for registered users. DEX2Web is backwards compatible with DEXi and can import models created by DEXi.

- *Registration* – This function allows the users of DEX2Web to register/create an account through a web browser. At this level, the input data required are in harmony with those required from the OAuth [18] internet protocol.
- *Log in* – This function allows users to log into the DEX2Web to use the functionality provided for registered users in the system.
- *Account editor* – This function allows users to edit all their provided account data given at the registration level.
- *Project editor* – This function allows DEX2Web users to create, edit, delete, import, export, and share a DEX project with other users. A DEX project can be saved on the DEX2Web server or on a local machine by exporting. An owner of a DEX project can add an unlimited number of users as members of that project.
- *User roles* – This function allows the users to specify a specific user's role in the case when they share a specific DEX project. Currently supported roles are *Decision Maker* and *Decision Analyst*.
- *Model editor* – This function allows users to create, modify, delete, import, export a DEX model, and share it with other users. The models supported in the current version have the same components as in DEXi, i.e., they are restricted to qualitative attributes and rule-based aggregation functions.
- *Attributes* – This function allows users to create, modify, and delete attributes. The concept of linked attributes in DEXi [10] is replaced by the more general concept of using full attribute hierarchies [9]. No restrictions are imposed upon the number of attributes included in a model. The largest DEX models constructed so far did not exceed 500 attributes [7], and this number is easily manageable by DEX2Web.
- *Scales* – This function allows users to create and modify scales of attributes. This version of DEX2Web supports categorical scales with any number of values, the same as DEXi.
- *Aggregation functions* – This function allows users to define and modify aggregation functions of attributes of type aggregated. Aggregation functions are currently represented only in the form of decision tables, the same as in DEXi. For practical reasons, the size of the tables is by default limited to the maximum of 2000 entries.
- *Values* – This function allows users to define DEX values to scales of attributes, the output of decision rules, and decision alternatives. The supported values are analogous to those in DEXi.
- *Alternatives* – This function allows users to define, modify, and delete alternatives.
- *Evaluation* – This function allows users to evaluate the defined alternatives. In this version of DEX2Web, only qualitative evaluation is supported.
- *Improved user-interface* – This function allows the users to see a DEXi-like view where one attribute from the model is selected, but also extends it with displays of properties and details of attributes, which are selectable by the user.
- *Editor of multiple DEX models* – This function allows users to edit up to five different DEX models simultaneously.
- *Search* – This function allows users to search in terms of DEX projects and DEX models by their names.
- *Charts* – This function allows users to build interactive charts. The charts supported by DEX2Web are of the type *Bar* and *Radar*.

- *Group decision-making* – This function provides support for group decision-making in two ways; (1) while sharing a DEX project with multiple users and (2) sharing a DEX model based on the Web Socket protocol [19]. Each user can define any number of models within a given project and share them with other users. There is no prescribed limit of the number of users sharing the same DEX model.

4.2 Non-functional Requirements

The non-functional requirements of DEX2Web are:

- *Usability*

  - The user interface is implemented using Thymeleaf.
  - The styling of web pages is done by Bootstrap front-end open-source toolkit.
  - Changes of the web pages' style are done in the Bootstrap toolkit's source files, and they are recompiled each time any change appears.

- *Reliability/Availability*

  - The software is deployed on a server that meets minimum requirements for a Spring project of version 2.3.4 that runs with Java 8.
  - The platform is designed to be available at all times, except in the rare cases of maintenance.
  - Before any update or upgrade of the platform, users are informed in advance through email and a notification on the platform's main page.

- Scalability

  - The whole platform project is built as a single JAR (Java ARchive).
  - The database is physically separated from the web-server.
  - The platform media are saved within the server.
  - The extension of volumes in storage space is done physically.

- Performance

  - The client requests are managed by thread connection pool.
  - The production database connections are managed by HikariCP [21] that supports the connection pool.
  - The port is 1000 Mbit/s with unlimited traffic.
  - The base of each webpage is light, and it downloads it under 250 ms.
  - Small AJAX requests are supported and allowed. The same is applied for web socket connections.
  - DEX2Web supports up to 1000 threads simultaneously.

- Serviceability

  - A specific documentation tool is developed within this project called *DexDoc*. This tool provides documentation of the platform DEX2Web and the DEX library. The DexDoc is interactive and can handle the next versions of both components.
  - The platform's logging is done using *Apache Commons Logging*, and for the DEX library, it is done from a built-in logging mechanism within the library.
  - The *Aspect-Oriented Programming (AOP)* is used to increase the modularity of cross-cutting concerns.

- Security

  - The server where DEX2Web is deployed is secured using network restrictions.
  - The web session timeout is set to 30 min.
  - HTTPS protocol is used to communicate between the browser and the server.
  - The validation of imported DEX files is done on two sides, user and server-side.
  - Nothing in the database is permanently deleted.
  - Database backup is done automatically every day.
  - Database backups are relocated manually to another secure server each week.
  - The authorization and authentication of the user are taken care of by following Spring Security [20].

4.3 Multi-layered Architecture

DEX2Web implements the DEX method using the newly developed DEX library. To fulfil the functional and non-functional requirements defined above, and to reduce the development time, we used a Java framework known as Spring Boot [20].

The architecture of DEX2Web is multi-layered due to the web-based nature of software. The software architecture is dependent on the architecture of the Spring Boot framework.

DEX2Web is a large-scale web-based enterprise application. DEX2Web application consists of two main parts: (1) resources and (2) business logic. The first group of resources consists of view layouts and media type resources such as images (158 files). The second group of DEX2Web application consists of Java classes (78 Java files) structured mainly following the *model-view-controller* design pattern.

DEX2Web implements a 4-layer architecture (see Fig. 1) consisting of:

- *Presentation layer*: Web pages for handling the dialogue with the user (decision maker) and invoking the necessary services.
- *Business layer*: It consists of the DEX library that provides means and methods for the creation and modification of DEX models, and evaluation and analysis of decision alternatives. Also, it includes all the other components that are mainly supported by the Spring Boot framework, such as handling the MVC design pattern and providing utilities needed to implement functional requirements.
- *Persistence layer*: It consists of components where we set up the communication channel with the database following *data-access-object* and *data-transfer-object* patterns.

- *Database layer*: A relational database designed using *MySQL*, which is used to store data and to facilitate sufficient information sharing between users and the library to provide full functionality of DEX2Web.

This implementation of DEX2Web is the first available online DMS that supports the DEX method for model development, and evaluation and analysis of alternatives. The software can be used remotely, which opens up new possibilities in decision support. To support group decision-making, such DMS requires a network to share information, and the Internet serves for this purpose. DEX2Web supports both individual and group decision-making.

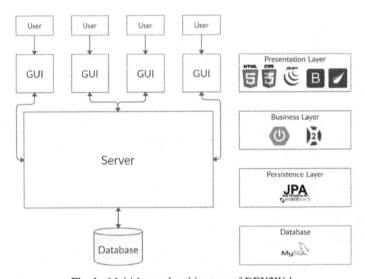

Fig. 1. Multi-layered architecture of DEX2Web.

4.4   Example of Using DEX2Web

This section presents a DEX2Web use-case, illustrating the development of DEX models and evaluation of decision alternatives. The example uses the CAR model, presented in Sect. 2.

Figure 2 shows the model editor of DEX2Web, which provides functionality for defining the hierarchical structure of the model, naming, and describing attributes, and accessing editors of other DEX model components (scales and aggregation functions). Figure 2 shows the structure of the CAR model in the stage where scales of attributes and aggregation functions have not been defined yet.

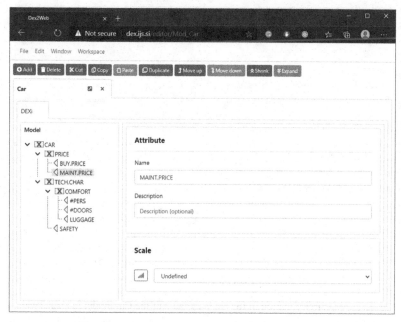

Fig. 2. Tree structure of the car model in DEX2Web.

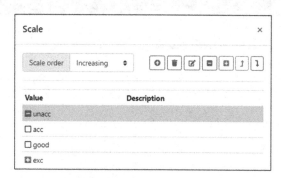

Fig. 3. Defining scale of the *CAR* attribute in DEX2Web.

Figure 3 shows the window used to define the scale of an attribute – specifically, the scale of the root attribute *CAR*, which consists of four preferentially-ordered qualitative values: *unacc, acc, good,* and *exc.*

The next step is to define the aggregation function of all aggregated attributes. Figure 4 shows the function editor while editing the decision table from Table 2. By default, output values are assigned to the value "*" (asterisk) that represents the set of all values of the scale of that attribute. Figure 4 shows the defined aggregation function of *CAR* aggregated attribute where the decision-maker has already defined the output values in the right-most column.

	PRICE	TECH.CHAR.	CAR
	Decision Rules CAR		✕
	Value unacc		⇕
1	high	bad	unacc
2	high	acc	unacc
3	high	good	unacc
4	high	exc	unacc
5	medium	bad	unacc
6	medium	acc	acc
7	medium	good	good
8	medium	exc	exc
9	low	bad	unacc
10	low	acc	good
11	low	good	exc
12	low	exc	exc

Fig. 4. Aggregation function of the *CAR* attribute of the Car model in DEX2Web.

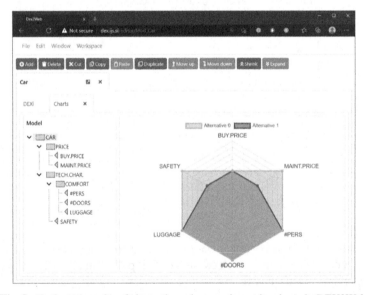

Fig. 5. Evaluated results of alternatives shown using radar charts in DEX2Web.

After defining all model components, the decision-maker can define decision alternatives and evaluate them. Figure 5 shows the results of two evaluated alternatives (cars) using radar charts.

Table 3. Functionality comparison between DEXi and DEX2Web.

Functionality	DEXi	DEX2Web
Online	✗	✓
DEX projects	✗	✓
Tabular view of DEX models	✗	✓
DEX attributes	✓	✓
DEX scales	✓	✓
DEX aggregation functions	✓	✓
Scale order in aggregation functions	✓	✗
Weight in aggregation functions	✓	✗
Reports	✓	✗
Editor of multiple DEX models	✓	✓
Multiple view creator	✗	✓
Interactive charts	✗	✓
Group decision-making	✗	✓

Table 3 presents a comparison of supported functionality of DEXi and DEX2Web. The latter is the first online DEX implementation and extends DEXi by supporting DEX projects (as collections of DEX models), provides additional interactive views of DEX models and charts, and supports some basic group decision-making features (sharing of models). At this stage, some advanced DEXi features (considering scale orders and attribute weights to aid aggregation function editing, advanced reports) are not implemented in DEX2Web, but will be gradually introduced in the future.

5   Conclusions

The purpose of this work was to design a new decision-making software that implements the DEX method in a web-based environment. The main goals were to define a modern software architecture and user interface, suitable for the web, and to utilize the newly developed DEX library, which is meant to provide an extended set of methods and tools for the new generation of DEX software.

DEX2Web is the first fairly complete implementation of the DEX method available for interactive use on the Internet (https://dex2web.ijs.si/). DEX2Web is backwards compatible with DEXi: it can import, but not export, DEXi models; the latter is due to the extended functionality of the DEX library, which is not available in DEXi. The first version of DEX2Web was completed in October 2020, so it is still a brand-new software without much-gained experience about its use in practice.

Currently, DEX2Web mainly resembles the functionality that is already available in the desktop software DEXi. In addition to user and safety management functions, which are necessary for the web environment, DEX2Web currently provides only minor extensions to DEXi: multiple models collected within DEX projects extended model-editing views and model sharing. However, the architecture of DEX2Web is flexible, and together with the new DEX library, additional features will be gradually added to DEX2Web. These include combining qualitative and quantitative attributes, using extended value types in aggregation functions and evaluation of alternatives (intervals, fuzzy and probability value distribution, and samples), multiple types of aggregation functions (e.g., for conversion between qualitative and quantitative attributes and weighted aggregation using weights), and additional support for group decision-making.

Acknowledgment. The authors acknowledge Jožef Stefan International Postgraduate School, where this work has been conducted, and the Public Scholarship, Development, Disability and Maintenance Fund of the Republic of Slovenia (Contract no. 11011-44/2017-14) that financially supported this research. The authors also acknowledge the financial support from the Slovenian Research Agency (research core funding No. P2-0103).

References

1. Bouyssou, D., et al.: Evaluation and Decision Models with Multiple Criteria: Stepping Stones for the Analyst, vol. 86. Springer, Boston (2006). 10.1007/0-387-31099-1
2. Greco, S., Figueira, J., Ehrgott, M.: Multiple Criteria Decision Analysis. Springer, New York (2016). 10.1007/978-1-4939-3094-4
3. Ishizaka, A., Nemery, P.: Multi-criteria Decision Analysis: Methods and Software. Wiley, Hoboken (2013)
4. Mustajoki, J., Marttunen, M.: Comparison of Multi-Criteria Decision Analytical Software-Searching for Ideas for Developing a New EIA-Specific Multi-Criteria Software (2013)
5. Bhargava, H.K., Power, D.J., Sun, D.: Progress in web-based decision support technologies. Decis. Support Syst. **43**(4), 1083–1095 (2007)
6. Power, D.J.: Web-based and model-driven decision support systems: concepts and issues. In: AMCIS 2000 Proceedings, p. 387 (2000)
7. Bohanec, M.: Multi-criteria DEX models: an overview and analysis. In: Stirn, L.Z., Borštar, M.K., Žerovnik, J., Drobne, S. (eds.) The 14th International Symposium on Operational Research in Slovenia. Slovenian Society Informatika–Section for Operational Research, Ljubljana (2017)
8. Bohanec, M., et al.: DEX methodology: three decades of qualitative multi-attribute modeling. Informatica **37**(1) (2013)
9. Trdin, N., Bohanec, M.: Extending the multi-criteria decision making method DEX with numeric attributes, value distributions and relational models. CEJOR **26**(1), 1–41 (2018)
10. Bohanec, M.: DEXi: program for multi-attribute decision making, user's manual, version 5.04. IJS Report DP-13100, Ljubljana, Slovenia: Institut Jozef Stefan (2020)
11. Bohanec, M., Rajkovič, V.: DEX: an expert system shell for decision support. Sistemica **1**(1), 145–157 (1990)
12. Kikaj, A., Bohanec, M.: Towards web-based decision modeling software based on DEX methodology. In: Central European Conference on Information and Intelligent Systems. Faculty of Organization and Informatics Varazdin (2019)
13. Kikaj, A., Bohanec, M.: Complex decision rules in DEX methodology: jRule algorithm and performance analysis. In: Proceedings of the 21th International Conference Information Society IS 2018, Volume A, Ljubljana, pp. 17–20 (2018)
14. Bohanec, M., Zupan, B.: Car Evaluation Dataset. UCI Machine Learning Repository. https://archive.ics.uci.edu/ml/datasets/car+evaluation. Accessed 25 Nov 2020

15. Bohanec, M., Rajkovič, V.: DECMAK: An Expert System Shell for Multi-Attribute Decision Making. Institut Jožef Stefan (1988)
16. Šet, A., Bohanec, M., Krisper, M.: VREDANA: Program za Vrednotenje in Analizo Variant v Večparametrskem Odločanju. Zbornik četrte Elektrotehniške in računalniške konference ERK, vol. 95 (1995)
17. Žnidaršič, M., Bohanec, M., Zupan, B.: proDEX–a DSS tool for environmental decision-making. Environ. Model. Softw. **21**, 1514–1516 (2006)
18. Bihis, C.: Mastering OAuth 2.0. Packt Publishing Ltd (2015)
19. Lombardi, A.: WebSocket: Lightweight Client-Server Communications. O'Reilly Media, Inc. (2015)
20. Spring Makes Java Simple. https://spring.io/. Accessed 25 Nov 2020
21. HikariCP. https://github.com/brettwooldridge/HikariCP. Accessed 25 Nov 2020

Comparison of AHP, PAPRICA, PROMETHEE, DEX and TOPSIS on an Application for Employee Selection

Anton Stipeč[1]([✉]) and Biljana Mileva Boshkoska[1,2]

[1] Faculty of Information Studies, 8000 Novo Mesto, Slovenia
[2] Institute Jožef Stefan, Ljubljana, Slovenia

Abstract. Employee selection is an essential process that in many organizations depends only on human judgement, and in today's fast-changing work environment, is susceptible to human errors. Hence, the problem of selecting the most appropriate employee is complex and costly, especially if the selected employee is unsuitable for the job position. The complexity of the problem arises from several requirements: each job position requires more than one criterion to be fulfilled by the job candidate; each candidate has a different set of skills, and usually, several candidates apply to one job position. The decision-maker has to make a quick selection decision, as the longer employee selection process is, the greater the costs are. In this article, we build 20 decision support models for four different job positions with five Multi-Criteria Decision Making (MCDM) methods and we compare them on a real set of data from an employment agency. The goal of this comparison is to recommend which method is most suitable by comparing the correctness of the results, ranking with missing values and difficulty to use. The results show which MCDM method is better for filtering most suitable employees given all required criteria and which MCDM method would be recommendable for employees ranking.

Keywords: Employee · Selection · Multi-criteria · Decision · Employment

1 Introduction

The employee selection in organisations is of utmost importance to survive in the increasingly demanding market. The role of selecting the best employee in a large organization is performed by the human resources (HR) department, however, small organisations lack time and experiences, hence they usually outsource this task to employment agencies.

In the process, the employment agencies face many potential candidates for one job opening while having a limited time for making the selection. Each candidate possesses certain skills, knowledge and personality, that need to be matched with the job position criteria leading to a complex problem. The main goal of the employee selection is hiring the best available employee, who will fulfil the required criteria and, in a certain way, be the right person for the job. If an employee is unsuitable or unhappy with the job position, wages, and/or working environment, his/her productivity will be decreased

© Springer Nature Switzerland AG 2021
U. Jayawickrama et al. (Eds.): ICDSST 2021, LNBIP 414, pp. 44–54, 2021.
https://doi.org/10.1007/978-3-030-73976-8_4

and that may result in more costs for the organization. The usual job criteria that are considered by the decision-maker are level and type of education, language, driver's license, previous job experience and other qualifications that depend on the job type (computer skills, professional driver's license, personal skills, etc.). The decision-maker may be further influenced by subjectivity and personal opinion. Only by examining all available information about each job candidate and rank the candidates, the decision-maker will be able to determine which one is the best for the job and finally make a decision. Since there is more than one criterion required to make the decision, employee selection can be defined as a Multiple Criteria Decision Making (MCDM) problem. To ensure that all gathered information about each job candidate is processed properly, to minimize the influence of decision maker's subjectivity and to decrease the time to select employees, an expert-based Decision Support System (DSS) based on an MCDM method, can be implemented. There are numerous MCDM methods available which can be used to improve employee selection, each has its advantages, disadvantages, strengths or weaknesses. In this paper, we compare five qualitative MCDM methods, TOPSIS, PROMETHEE, PAPRICA, DEX and AHP to determine their strengths, weaknesses and differences thus allowing researchers to select the most suitable one for the preparation of a DSS for employee selection or other similar MCDM problem.

The paper is organized as follows. In Sect. 2 we present the related work of other MCDM methods applied to the problem of the employee selection. In Sect. 3 we describe a case study used for comparison of the MCDM methods. In Sect. 4 we describe the MCDM methods used in this research. In Sect. 5 we present and discuss the ranking results and in Sect. 6 we present the conclusion of this research.

2 Related Work

Employee selection has been extensively researched by many authors. In his article, Acikgoz [1] described an integrative model of job search and employee recruitment, which can help explain specific relationships between individuals and organization in the recruitment process, how the organization approaches individuals and explain the individual's approaches to the organization in the recruitment process. Kalugina & Shvydun [2], in their article, dealt with the problem of employee turnover as a major issue for today's companies. They presented a model for matching candidates and employers in the process of employee selection which then could be used to ensure that all relevant information is covered to select the best employee who is qualified for the job and compatible with the organization to avoid turnovers as much as possible. The problem of overconfidence in employee selection was addressed in an article written by Kausel, Culbertson, & Madrid [3]. Three studies were carried out and results showed that in the case of unstructured interviews and lack of correct and precise information about the job candidate, the employee selection depends on the personal impression of an interviewer, which can result in an error in the assessment of candidates hence they should be avoided if possible. Alqahtani et al. [4] propose fuzzy logic-based E-Recruitment DSS to help decision-makers overcome the limitations of existing recruitment processes, who in most cases, select candidates who meet 100% of the criteria and turns down all candidates who partially meet the criteria.

Many authors have considered using MCDM methods to help make better decisions in employee selection. TOPSIS was applied, in combination with the AHP, to select a manager of the telecommunication company in Indonesia [5]. Siregar et al. [6] have proposed a DSS for assessment of job candidates based on TOPSIS method and Aarushi [7] proposed a methodology which used AHP to rank job criteria and TOPSIS to rank candidates. PROMETHEE method was applied in combination with the AHP method to select candidates for the election in the Grand National Assembly of Turkey [8]. Combination of AHP-PROMETHEE methods was applied in the process of reorganization and downsizing of the company [9]. DEX method was used for employee redeployment in large governmental organizations [10]. AHP method is used to select a director of a geomatics laboratory [11] and in another problem, to determine the factor that has the most influence on the organization's work climate [12]. Yavuz and Gokhan used AHP to select employees for promotion [13]. 1000 minds software [14], which is based on the PAPRICA method, was applied to a problem of awarding student scholarships at the University of Otago in New Zealand and for allocating research grants at the University of Otago and for selection of Teaching Fellows in the Department of Design Studies. Some of these methods were also applied to solve other types of MCDM problems. The PAPRICA method was applied to a problem of a medical device selection for a University hospital [15], or on a problem of a cloud service selection [16]. TOPSIS and PROMETHEE methods were compared together with VIšeKriterijumska Optimizacija I Kompromisno Rešenje (VIKOR) and ELECTRE on a problem of selection of materials for an automotive instrument panel [17]. TOPSIS, PROMETHEE, AHP, ELECTRE and Weighted Sum Model (WSM) method on a problem of urban water infrastructure [18].

We considered TOPSIS, PROMETHEE, PAPRICA, DEX and AHP methods for research because they are widely used to solve various MCDM problems and to the knowledge of the authors they have not been compared specifically to the employee selection problem.

3   A Case Study for Comparison of MCDM Methods

Five MCDM methods are used to rank fifteen job candidates to four specific and diverse job positions: truck driver, accountant, auxiliary worker and project manager. Job positions and required criteria are specified in Table 1. Job positions were specifically selected as diverse as possible to determine the goodness of ranking of each MCDM method to the baseline ranking provided by the decision-maker. The criteria that are used to rank candidates are education type and level, foreign language type and level, driver's license category, computer skills, additional qualifications, communication skills (communication, listening, presenting), organizational skills (planning, organization, decision making, supervision), leadership skills (management and delegation of responsibility), work skills (flexibility, responsibility, work initiative, type and duration of work experience). Each selected candidate possesses specific skills and each job position requires some or all of the criteria to be fulfilled by the job candidate. The candidate who has a better overall score should be ranked above the one with a lower score, leading to an overall ranking of candidates. The problem that occurs is that each method ranks candidates slightly differently. Hence we need to propose a metric of how to choose among different MCDM methods.

Table 1. Job positions and the required criteria

Criteria	Job position			
	Truck driver	Accountant	Auxiliary worker	Project manager
Education type	Professional driver	Economy	–	Economy
Education level	IV	IV	–	VII
Language type	–	–	–	English
Language level	–	–	–	C1
Driver's license	C	–	–	B
Computer skills	–	Basic	–	Basic
Additional qualifications	Code 95	–	–	–
Communication	Good	–	–	Excellent
Listening	Good	–	–	Excellent
Presenting	–	–	–	Excellent
Planning	Good	Good	–	Excellent
Organization	Good	Good	–	Excellent
Decision making	–	–	–	Excellent
Supervision	–	–	–	Excellent
Management	–	–	–	Excellent
Delegation of responsibility	–	–	–	Excellent
Flexibility	Good	–	Good	Excellent
Responsibility	Good	Good	Good	Excellent
Initiative	Good	–	Good	Excellent
Experience type	Professional driver	Accountant	Any	Project manager
Experience duration (in years)	1	–	1	2

The 15 chosen candidates have diverse education types. Five candidates are "Professional driver", three are "Economist" and the rest are "Traffic engineer", "Traffic technician", "Carpenter", Car mechanic" and "Hairdresser". Majority of the candidates have IV level of education, one has VI level and two have VII level of education. Also, many of the candidates stated that speak English as a foreign language, while some speak Italian, German or Russian. Candidates that have "Professional driver" education type have "Code 95" additional qualification and drivers license C category, others have "Taxi driver certificate" or "Forklift operator". Several candidates have B category drivers license. Regarding computer skills, eight candidates have "Basic" and three have "Advanced" computer skills while all the rest have none. Twelve candidates stated to

have "Good" communication and listening skills, eight candidates are "Good" at presenting skills, three candidates have "Excellent" communication, listening and presenting
skills, and the rest have "Bad" for all categories. Eight candidates have "Good" planning skills, one has "Excellent" and others have "Bad". Ten candidates have "Good"
organizational skills and four candidates have "Bad". Four candidates have "Good"
decision-making skills while all others have "Bad". Nearly all candidates have "Bad"
supervision, management and delegation of responsibility skills, except two that have
"Good" supervision skills and one have "Good" management and delegation of responsibility. Eight candidates have "Good" flexibility, seven have "Excellent", ten candidates
have "Good" responsibility and initiative, others have "Excellent". Experience type is
diverse, five candidates have "Professional driver", two have "Accountant" others have
"Mechanic", "Baker" or "Auxiliary worker". Experience duration varies from six months
to six years.

4 Description of the MCDM Methods

Analytic Hierarchy Process (AHP) is a multi-criteria decision-making method which
was developed by Saaty [19] in 1971. AHP does not prescribe the correct decision
for the problem, but instead, AHP helps decision-makers find the best decision that
suits their goal. Using AHP decision-makers build a hierarchy by decomposing the
decision problem into easily comprehended sub-problems which then can be analyzed
independently. After hierarchy is built decision-makers can evaluate its elements by
comparing them to each other two at a time, considering their impact on element above
in the hierarchy. The results of these evaluations are numerical values which represent
weight value of each criterion, or its importance to criteria above. Decision-makers
can use concrete data if available, but in most cases, they use personal judgement in
determining the weight. This method can be used in the ranking of any set of variables.
The hierarchy of the AHP process is constructed with decision alternatives at the bottom
level, decision criteria in the middle level and decision goal at the top-level. Since there
are 15 alternatives, absolute measures will be used with AHP to decrease the number of
pairwise comparisons, which means that all 15 alternatives are not directly compared,
instead, categories are set for each criterion and these categories are pairwise compared
to each criterion individually. After the categories are defined and compared they are
inputted for each alternative depending on how that alternative meets each criterion.
"SuperDecisions" software was used to rank candidates using AHP.

TOPSIS method is one of the multi-criteria decision-making methods first introduced
by Yoon and Hwang [20] which uses a principle that the alternatives chosen must have the
shortest distance from a positive ideal solution and farthest distance from a negative ideal
solution. "The positive ideal solution is defined as the sum of all the best value that can
be achieved for each attribute, while the negative ideal solution consists of all the worst
value obtained for each attribute" [6, p. 7]. Advantage of using the TOPSIS method is that
it limits the subjectivity of decision-maker. Decision-maker only determines the weights
of the criteria. Other benefits of TOPSIS method are that it includes logic that represents
a rational human choice, it has a simple calculation procedure that is easy to program
and the results obtained for all the solutions can be visualized by polyhedron for any two

dimensions. TOPSIS method also has disadvantages. One of the main disadvantages is that it derives the weights and checks the consistency of decision-maker [21]. "Microsoft Excel" was used to rank candidates using TOPSIS.

PROMETHEE method was first developed by Brans in 1982. [22]. And extended by Brans and Vincke in 1985. [23]. PROMETHEE compares the alternatives considering the deviations of alternatives to each criterion. PROMETHEE uses positive and negative preference flows for each alternative to generate the ranking. PROMETHEE method can be applied to the variables included in the decision matrix without any normalization and is applicable when there is missing information. There are different versions of PROMETHEE: PROMETHEE I – partial ranking, PROMETHEE 2 – complete ranking, PROMETHEE III – a ranking based on intervals, PROMETHEE IV – continuous case, PROMETHEE GAIA – geometrical analysis for interactive assistance, PROMETHEE V - MCDA including segmentation constraints, and PROMETHEE VI - representation of the human brain. "Visual PROMETHEE" software was used to rank candidates using PROMETHEE.

The PAPRICA method, similar to AHP, is based on a pairwise comparison of the alternatives, in this case, medical devices [24]. Each decision-maker is repeatedly presented with the pair of alternatives in random order to choose which alternative he prefers. Each time two alternatives are ranked, other alternatives that can be ranked via transitivity are identified and eliminated. For example, if decision maker prefers alternative 1 over alternative 2, then he prefers alternative 2 over alternative 3. Advantage of the PAPRICA method is that it resembles human thinking. It is much easier to make a choice when there are few, in this case, a pair of alternatives to choose from. "1000 Minds" software was used to rank candidates using PAPRICA.

DEX method was developed by Rajkovič and implemented in DEXi software tool by Bohanec [25]. It is based on a hierarchical structure or hierarchy tree of the problem [26]. To obtain an overall aggregation, decision alternatives are first evaluated by input and aggregated attributes. Overall aggregation serves for the ranking, comparison, analysis and selection of alternatives. In the case of employee selection, overall aggregation represents employee which is decomposed into several qualifications (education, experience, language), these are further decomposed into attributes of each qualification, for example (language: English, German, etc.). For each available job, the rule table is defined, which DEX model uses to determine if a job candidate's qualification is excellent, good or bad for a specific job. These grades of the qualifications are then aggregated and, again by using the defined rule table, overall aggregation grade is determined, employees are ranked by overall grade. Python was used to rank candidates with Qualitative-Quantitative (DEX/QQ) method [27].

5   Results and Discussion

We compared the five MCDM methods based on the three criteria: (I) the similarity and the correctness of the results, (II) difficulty to use and (III) dealing with the missing values.

The similarity of rankings between each pair of the methods including the baseline ranking is measured using Kendall's tau rank correlation coefficient for each job position as shown in Table 2. Each square of the heatmap represents correlation rank for the compared two rankings obtained either by MCDM methods or baseline ranking. For the "Truck driver" job position the highest similarity is obtained between PAPRICA and both TOPSIS and PROMETHEE (0,88), while the highest similarity to the baseline ranking is obtained with TOPSIS (0,7). For the "Accountant" job position, the highest Kendall's tau rank correlation coefficient is obtained between PAPRICA and the baseline ranking (0,8). DEX show the least similarity to the baseline ranking (−0,22). For the "Auxiliary worker" job position the most similar rankings have TOPSIS and PROMETHEE (0,98) and DEX and PAPRICA have the most similar ranking compared to the baseline ranking (0,79). Most dissimilar rankings have PROMETHEE to the baseline ranking (0,71). For the "Project manager" job position the most similar rankings are obtained between TOPSIS and the baseline ranking (0,62). Most dissimilar rankings have DEX compared to the baseline ranking (−0,067).

Table 2. HeatMap of Kendall's tau correlation coefficient for the truck driver, Accountant, auxiliary worker and project manager job position

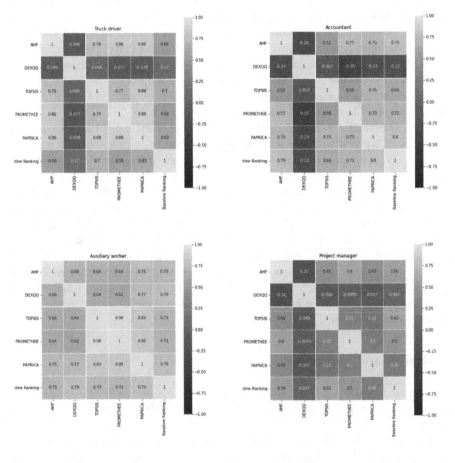

Given the differences between the rankings, we conclude that the rankings depend on the similarity between the job and the model. This results due to differences in understanding between an employer who seeks good candidates and the worker in the employment agency. One possible way to avoid such situations is to prepare different models for each job position.

The measurement of the correctness of the MCDM rankings, compared to the baseline rank, is displayed in Table 3, for each job position. Metric is defined with three values, where the number 3 denotes the acceptable ranking which matches the baseline rank for all candidates who meet the required criteria and these candidates are ranked above candidates who do not, 2 denotes for the rankings that have some differences in the ranking order when compared to the baseline ranking, however, the expected candidates are ranked higher than those who do not fulfil all criteria. The number 1 denotes rankings where candidates who do not fulfil all criteria are ranked higher than those who do. Models built using the five MCDM methods that resemble the closest ranking to the baseline rank are: DEX and TOPSIS for the "Truck driver" job position, AHP and DEX for the "Accountant" job position, TOPSIS and PROMETHEE for the "Auxiliary worker" job position and TOPSIS for the "Project manager" job position. For the "Project manager, because all of the candidates do not meet the required criteria," correctness was measured by how much ranking resembles the baseline rank.

Table 3. Measurement of the correctness of the MCDM rankings

	Truck driver	Accountant	Auxiliary worker	Project manager
AHP	3	2	2	2
DEX	3	3	2	1
TOPSIS	3	1	2	3
PROMETHEE	1	1	2	2
PAPRICA	2	3	2	2

Regarding the difficulty to use, each method has its advantages and disadvantages. Both AHP and PAPRICA methods require a large amount of input from the decision-maker. When using AHP, the decision-maker must first input preference values for every pair of criteria to determine criteria weights and then, input preference values for every pair of candidates for every criterion individually. These values determine which candidate is more preferred or which candidate better fulfils certain criterion. This means there are about 50 to 70 pairs for criteria analysis and more than 100 pairs of alternatives for 15 candidates. The number of pairwise comparisons depends on the number of criteria and available. The PAPRICA method requires a smaller amount of work than AHP method. The number of pairwise comparisons does not increase when the number of alternatives and criteria increase because the PAPRICA ranks via transitivity, as the decision-maker inputs preference values for a pair of criteria, these values are considered for further pairwise comparisons. Regarding the amount of work required from the decision-maker, DEX method requires that the decision-maker defines rule tables for each aggregation of the criteria in the hierarchy. Using rule tables, DEX is capable to quickly rank any amount of alternatives. TOPSIS and PROMETHEE methods require minimal input from the decision-maker. One problem is that both methods are suitable only for quantitative data, decision-maker must convert qualitative data to quantitative ones to rank alternatives. For example: bad = 0, good = 1, excellent = 2 and so on. Hence, additional transformation is required from qualitative to quantitative space. The number of calculations does not increase with the number of criteria and alternatives. One advantage of the PROMETHEE method over TOPSIS is that it does not require normalization.

When missing values are considered, DEX method ranked the alternatives correctly, but for alternatives which do not meet some of the criteria, DEX gave no information about the differences between criteria fulfilment. Other used MCDM methods could provide information about which candidate is better than the other even if candidates do not fulfil the required criteria. The best example is "Project manager" where all candidates do not fulfil required criteria and DEX wasn't capable to rank them, and it requires additional methods, such as QQ (Qualitative-Quantitative) method to solve the ranking issue.

6 Conclusion

This analysis showed that every MCDM method has its advantages and disadvantages which may result in differences in the rankings. The main goal is to find an MCDM method that ranks candidates as correctly as possible with minimal input required from the decision-maker. DEX ranked candidates exactly as the baseline rank for the "Truck driver" and "Accountant" job positions.. The main drawback with DEXi is that it does not include a tool for differentiating among candidates that are evaluated into the same evaluation group. AHP proved to be the most dependent on the decision-makers personal preference regarding the definition of the categories for each criterion and pairwise comparison between categories. On the other hand, AHP in most cases ranked candidates with small differences to the baseline rank. A similar situation is with the PAPRICA method, but it requires less amount of input from the decision-maker. AHP is best suitable when there are less than ten alternatives (unless absolute measures are used), while when using other four MCDM methods, the amount of input does not increase when alternatives are added. PROMETHEE and TOPSIS require minimal input from the decision-maker, but there are differences in the results. For the "Truck driver" and "Accountant" job positions where some of the candidates fulfil all required criteria, PROMETHEE ranked candidates who do not fulfil all criteria above those who do. PROMETHEE does not require normalization but it is difficult to keep an overview when a large number of criteria is used, and the criteria compensation. TOPSIS placed candidate 15, who do not have required experience type and duration, in the first place for the "Accountant" job position because of the criteria compensation, it requires normalization and both TOPSIS and PROMETHEE are suitable only for quantitative data. Meaning that qualitative data, such as education type, or experience type must be converted to quantitative before rankings are made. As results showed, the correctness of the rankings depends on the job position. Since rankings obtained by AHP and PAPRICA depend on the personal input of the decision-maker and rankings obtained by TOPSIS and PROMETHEE on the conversion of qualitative data to quantitative and on the calculation of the results, DEX proved to be the most straight forward Considering all said, DEX could be used to rank and filter candidates who fulfil all required criteria and PROMETHEE or TOPSIS could be used for additional ranking of filtered candidates since all candidates who do not fulfil all criteria are filtered out.

Acknowledgement. The second author acknowledges funding from the Slovenian Research Agency via program Complex Networks P1-0383 and the funding received from the European Union's Horizon 2020 research and innovation programme project HECAT under grant agreement No. 870702.

References

1. Acikgoz, Y.: Employee recruitment and job search: towards a multi-level integration. Hum. Resour. Manag. Rev. **29**(1), 1–13 (2018)
2. Kalugina, E., Shvydun, S.: An effective personnel selection model. Procedia Comput. Sci. **31**, 1102–1106 (2014)
3. Kausel, E.E., Culbertson, S.S., Madrid, H.P.: Overconfidence in personnel selection: when and why unstructured interview information can hurt hiring decisions. Organ. Behav. Hum. Decis. Process. **137**, 27–44 (2016)
4. Alqahtani, A., Alhaidari, F.A., Rahman, A., Mahmud, M., Sultan, K.: Decision support system assisted e-recruiting system. J. Comput. Theor. Nanosci. **16**, 335–340 (2019)
5. Kusumawardani, R.P., Agintiara, M.: Application of fuzzy AHP-TOPSIS method for decision making in human resource manager selection process. Procedia Comput. Sci. **72**, 638–646 (2015)
6. Siregar, D., Tinggi, S., Harapan, T.: Decision support system best employee assessments with technique for order of preference by similarity to ideal solution. Int. J. Recent Trends Eng. Res. **3**, 6–17 (2017)
7. Aarushi, S.M.: Generalized MCDM-based decision. In: Advances in Intelligent Systems, pp. 155–167 (2016)
8. Kazan, H., Özçelik, S., Hobikoğlu, E.H.: Election of deputy candidates for nomination with AHP-promethee methods. Procedia Soc. Behav. Sci. **195**, 603–613 (2015)
9. Bogdanović, D., Miletić, S.: Personnel evaluation and selection by multicriteria decision making method. Econ. Comput. Econ. Cybern. Stud. Res. **48**, 22–39 (2014)
10. Hajnić, M., Boshkoska, B.M.: A decision support model for the operational management of employee redeployment in large governmental organisations. J. Decis. Syst. 1–9 (2020)
11. D'Urso, M., Masi, D.: Multi-criteria decision-making methods and their applications for human resources. In: ISPRS - International Archives of the Photogrammetry, Remote Sensing and Spatial Information Sciences, pp. 31–37 (2015)
12. Phudphad, K., Watanapa, B., Krathu, W., Funilkul, S.: Rankings of the security factors of human resources information system (HRIS) influencing the open climate of work: using analytic hierarchy process (AHP). Procedia Comput. Sci. **111**, 287–293 (2017)
13. Yavuz, O., Kemal Gokhan, N.: Personnel selection for promotion using an integrated consistent fuzzy preference relations - fuzzy analytic hierarchy process methodology: a real case study. Asian J. Interdisc. Res. **3**, 219–236 (2020)
14. 1000 minds. https://www.1000minds.com/sectors/government/scholarships. Accessed 5 Sept 2018

15. Martelli, N., et al.: Combining multi-criteria decision analysis and mini-health technology assessment: a funding decision-support tool for medical devices in a university hospital setting. J. Biomed. Inform. **59**, 201–208 (2016)
16. Alismaili, S., Li, M., Shen, J., He, Q.: A consumer-oriented decision-making approach for selecting the cloud storage service: from PAPRIKA perspective. In: Fan, M., Heikkilä, J., Li, H., Shaw, M.J., Zhang, H. (eds.) WEB 2016. LNBIP, vol. 296, pp. 1–12. Springer, Cham (2017). https://doi.org/10.1007/978-3-319-69644-7_1
17. Gul, M., Celik, E., Gumus, A.T., Guneri, A.F.: A fuzzy logic based PROMETHEE method for material selection problems. Beni-Suef Univ. J. Basic Appl. Sci. **7**, 68–79 (2017)
18. Tscheikner-Gratl, F., Egger, P., Rauch, W., Kleidorfer, M.: Comparison of multi-criteria decision support methods for integrated rehabilitation prioritization. Water **9**(2), 68 (2017)
19. Saaty, T.: The Analytic Hierarchy Process. McGraw-Hill, New York (1980)
20. Hwang, C., Yoon, K.: Multiple Attribute Decision Making: Methods and Applications. Springer, New York (1981). https://doi.org/10.1007/978-3-642-48318-9
21. Łatuszyńska, A.: Multiple-criteria decision analysis using topsis method for interval data in research into the level of information society development. Folia Oeconomica Stetinensia **13**(2), 63–76 (2014)
22. Brans, J.P.: Elaboration d'instruments d'aide à la décision, La méthode PROMETHEE. In: Nature, Instruments et Perspectives d'Avenir, pp. 183–213 (1982)
23. Brans, J.P., Vincle, P.: A preference ranking organization method. Manag. Sci. **31**, 647–656 (1985)
24. Kabir, G., Sadiq, R., Tesfamariam, S.: A review of multi-criteria decision-making methods for infrastructure management. Struct. Infrastruct. Eng. **10**, 1176–1210 (2014)
25. Bohanec, M., Boshkoska, B.M., Prins, T., Kok, E.J.: SIGMO: a decision support System for Identification of genetically modified food or feed products. Food Control **71**, 168–177 (2017)
26. Bohanec, M., Rajkovič, V., Bratko, I., Zupan, B., Žnidaršič, M.: DEX methodology: three decades of qualitative multi-attribute modelling. Informatica **37**, 49–54 (2013)
27. Damij, N., Boškoski, P., Bohanec, M., Mileva Boshkoska, B.: Ranking of business process simulation software tools with DEX/QQ hierarchical decision model. PLoS ONE **11**, 16 (2016)

A Survey on Criteria for Smart Home Systems with Integration into the Analytic Hierarchy Process

Georg Wieland$^{(\boxtimes)}$ and Herwig Zeiner

JOANNEUM RESEARCH Forschungsgesellschaft mbH,
Steyrergasse 17, 8010 Graz, Austria
{georg.wieland,herwig.zeiner}@joanneum.at
http://www.joanneum.at

Abstract. While smart home systems and smart home applications are utilized more and more, the assortment of smart products gets broaden and is accompanied by a constant growth. This stable increase of products leads to confusion and perplexity and, consequently, it hinders a deliberate and well thought through decision. This paper presents a survey on the most important criteria which enable a conscious assessment of smart home systems. It provides an elaborate review of various research outputs pertaining usability, sustainability and complexity of smart home applications. By connecting such applications with the terms platforms and IoT, a new door is opened enabling an up to date assessment. Security, safety as well as data protection extend the discussion and shepherd towards a most complete description in order to appraise smart home products. Finally, based on the reviewed topics, a list containing 18 discretized criteria is given which allows an integration into the Analytic Hierarchy Process.

Keywords: Smart home · Domestic technology · Literature review · Assessment · Analytic hierarchy process

1 Introduction

1.1 Motivation

Over the last twenty years smart home systems (SHS) and smart home applications (SHA) [28,32] gained more importance as well as acceptance in the user and developer community. This increase in attention and awareness can be briefly explained by new emerged possibilities due to progressive development regarding hardware and software components. Especially the ongoing evolutionary progress in the fields of Internet of Things (IoT) [22] and web-based applications [17] contributes to a constant growth of opportunities which broaden the usability in highly diverse use cases and research areas. In virtue of the reasons mentioned

© Springer Nature Switzerland AG 2021
U. Jayawickrama et al. (Eds.): ICDSST 2021, LNBIP 414, pp. 55–66, 2021.
https://doi.org/10.1007/978-3-030-73976-8_5

above, SHS enable an extensive application in private as well as commercially utilized premises. Also, such systems are applicable in various fields of research (e.g. smart sensing technologies, energy-aware buildings, security and safety, ambient assisted living (AAL)).

Before immersing into this subject, a lucid as well as applicable definition of SHS and SHA is required. Drawn from the description given in [23], SHS are classified as frameworks which enable monitoring and controlling processes for domestic attributes including all needed parts, such as sensors, actuators, control devices, software, etc. Similarly, SHA describe underlying regimes but not necessarily all involved physical parts, because some components might not be included in the offered product.

Given a comprehensive range of locally and web-based SHS, a difficulty arises when being after an adequate and compatible system for specific necessities. In order to determine a suitable SHS for personal or research purposes, it is necessary to review miscellaneous systems and applications in greater detail. Therefore, one may consider criteria which facilitate this process and serve as a generic guideline for a precise classification and evaluation of SHA. This paper presents a general collection of well researched criteria with respect to divergent view points and different segments. Hence, it serves as a generic guidance while assessing SHS and, likewise, gleans relevant information about the standards of such systems.

The first part of this publication, Sect. 2, states the generic criteria for SHS. In this context, the terms usability, sustainability, complexity and user-adaptation are going to be discussed. Furthermore, platform-concepts and IoT as well as safety and privacy aspects are examined. Section 3, extents the discussed criteria into a usable list which serves as a general tool to apply the Analytic Hierarchy Process in order to asses SHS and SHA.

1.2 Related Work

SHS and their field of research is well established in the landscape of smart technologies. For Example, Alam et al. [3] present an overview of previous research in this area of analysis, discussing building blocks and interrelationships pertaining these components and also reviewing algorithms as well as communication protocols. Furthermore, the article in [2] brings additional options and gaps respecting this line of research to the table. It represents a consistent piece of work including SHS, different smart applications and IoT. On the contrary, numerous research outputs are framed narrowly, centered around a particular topic describing applications specifically, for example see [27]. This fact, moreover, corresponds to the apparent growth of the smart home market as well as the expanding field of research. Such precise publications, though, enable extensive surveys referring SHS, which, eventually, summarize the state of art and give a complete insight on the topic. Likewise, the creation of research content combining SHS and IoT is absolutely current.

Similar work to this paper regarding general criteria for SHS is, for example, given in [7]. Section 2 of this paper presents a short overview over various requirements leading to a realization using a device based web-service called WS4D-PipesBox[1]. Another example can be found in [13]. This work of research reviews different criteria which are important for the designing process of acceptable smart home technologies. These enumerations of criteria for SHS, however, are constructed to discuss certain applications or further proceedings.

Therefore, the work presented in this paper is focused on establishing a generic set of criteria enabling a general assessment of existing SHS and SHA. Compressing this set into a discrete list facilitates the usage of the Analytic Hierarchy Process in order to make decisions on the ideal smart home product deliberately. This approach enables research groups, developers, builders as well as users to find the ideal product for their requirements. This serves as the main scientific contribution of this paper.

2 Smart Home Criteria

The aim of smart home products is to redefine the domestic as well as the professional life by providing support and assistance for everyday procedures to enhance the quality of living and, moreover, to increase workflow and productivity. However, this goal implies the usage of an aligned SHS which fulfills the needs and requirements of the user. Due to the large amount of existing smart home products and, therefore, nearly endless possibilities, an assessment of various SHS and SHA is needed. This also holds, without proof, for research purposes. Before discussing criteria for SHS, it is indispensable to note that these criteria should only give an overview of relevant topics by observing the current stage of research. As mentioned above, a generic guideline for assessing SHS and SHA is going to be presented.

2.1 Usability and Sustainability

Foremost, usability and sustainability regarding SHS cannot be observed separately. The applicability depends on functionality, durability and trouble-free processing which implies a conjunction of the terms mentioned in the title of this subsection.

General Features. SHS have to perform appropriately respecting the users premises. Such systems also need to satisfy the conditions of their surroundings as well as their field of application. Due to the headway of technology, almost no compromises regarding general features have to be made. However, other circumstances, discussed below, restrict the capabilities of SHA. For example, let us behold the area of health care equipped with the term "smart", proposed by [10]. Existing SHA, such as fall detection or smart heart monitoring, are able to enhance the health care sector and, especially, the area of AAL. Suchlike technologies can assist, e.g., elderly people living at home as well as hospital personnel

---

[1] http://ws4d.org/software-downloads/pipesbox/.

monitoring vital parameters of patients. Nevertheless, this yields various challenges concerning safety and privacy which have to be overcome. Precautions regarding security issues, as another example, can also be targeted via SHS, see e.g. [4]. Smoke detectors, burglary alarms, surge probes, et cetera could be included in order to prevent and protect. However, these provisions forbid errors of the used system which restrain the functionality and therefore, a sustainable and reliable SHS is needed.

Controlling and Visualization. Lucid control options [29] and a clear representation of ongoing processes [33] are indispensable necessities of SHS. In regards of controlling and illustrating the system via local units, suitable locations, accessible to all users, as well as appropriate regulation and depiction options have to be chosen. The governing questions are: What should be displayed? What should be controllable? These questions also hold for the usage of web-based services and the utilization of smart phones. Here, an obvious advantage is that every user can adjust and monitor the system at all time on the go. However, due to the individuality of different users, adaptable apps and services are required.

Added Values. Obviously, the aim of SHS is to add certain values to the domestic life and/or work performance of the user. In [18] fieldwork regarding these appended values is done. Apparently, new emerged opportunities and possibilities originating from developments in the smart home area yield additional values for the users home or work place. However, these values should be gauged accurately in order to weigh the benefits of a SHS appropriately.

User Investments. With respect to sustainability, user investments of money and time have to be assessed. On the one hand, (half-)closed systems are often sold with broad installment and service offers which imply a reasonable investment of time. Nevertheless, such systems are regularly found in a higher prize range. Thus, large investments in the smart home sector correspond to less needed knowledge and time installing and operating a SHS. On the other hand, existing free (open) SHS, such as openHab[2] or ioBroker[3], cost less money due to the fact that they can be paired with standard smart products, for example with the Philips hue line[4]. Such constructed systems, though, require previous knowledge as well as the will to invest time because the entire chain of "smart" elements has to be structured by the user. [25] discusses technical barriers and general challenges implementing SHA.

Updates and Improvements. With the purpose of choosing a sustainable SHS, updates and improvements should be available as well as feasible. This

---

[2] https://www.openhab.org/.

[3] https://www.iobroker.net/.

[4] https://www.philips-hue.com/.

introduces new challenges achieving sustainability [37] which have to be assessed beforehand. Updates, in general, shall advance the system and, also, ensure a functioning interaction between all components. Improvements, however, should be possible in order to add new features and elements to the system since technology extends consistently.

Energy-Awareness and Ecological Footprint. Lastly, in conjunction with sustainability the concept of a energy-aware and ecological sensible SHS, discussed in [5], should be reviewed. Resources are limited and therefore, they have to be utilized cautiously. This pertains the development and usage of generic technologies and hence, energy-awareness and ecological sensibility affects SHS in various ways. For example, hardware parts, such as sensors and switches, have to be produced sustainably as well as durable and long-lasting. Moreover, energy consumption, especially current drain, should be taken into consideration.

2.2 Complexity and Integration

The complexity of SHS can be assessed independently but also in conjunction with further integration, which describes the process of adding new components and features to the system. To demonstrate complexity of SHS by itself see, for example, [8] which reviews different sophisticated learning models for smart homes.

Setup, Usage and Trouble Shooting. This section holds primarily for open SHS because the application of these systems requires prior technical and computational knowledge. Despite the fact that manuals and instructions can be found online, the realization comes with difficulties to non-professional users. [9], for example, describes these problems and challenges in greater detail. The main difficulties arise while implementing a correct setup. This procedure takes time and previous knowledge because all the necessary elements of a SHS have to be taken into account. Therefore, the users objectives have to be distinct beforehand and based on these intentions a respective selection regarding e.g. sensors, controls, visualization, communication as well as storage has to be made. However, manually setup open systems simplify trouble shooting and the usage of the system itself due to the gained knowledge while accomplishing the setup. Apart from that, (half-)closed SHS are mostly setup prior which reduces the complexity. Nevertheless, one has to learn the system and this could, respectively, increase complexity; trouble shooting could be covered by the distributor but could also lie in the hands of the user.

Custom Features and Additional Integration. In case a SHS user has the desire to customize its features, possibilities to do so should be available. By that means, one could program custom rules which connect predefined parameters with certain consequences enabling automation and immediate changes of the system. Besides that, a conjunction of different features could be possible. This, however, boosts complexity and therefore a lucid medium for custom adjustments is needed. Additional integration [21] of sensors and actuators improves the SHS and enables further customization. For open, manually setup systems this could be fairly trivial due to the acquired knowledge and the awareness of the system. (Half-)closed SHS mostly do not support self-integration of additional elements. However, these products often endorse various companies and their "smart" devices. In this case, further integration is primarily automatized.

2.3   User-Adaptation

As the user changes, a SHS has to change as well and therefore, appropriate adaptations are desirable. These adaptations should guarantee a valuable usage of the system for different groups of people - the users. User-adaptation can be achieved by implementing learning techniques to analyse reoccurring patters of various users with which the system can adjust itself respectively and generate suitable automation control parameters and options, for example see [36]. At this point, it is also important to discuss the acceptance of SHS in society. In order to incorporate SHA in everyday life, all affected people have to grapple with changes and new opportunities. Thus, there are sociological challenges [16] regarding SHS that have to be overcome. Also, ethical issues [12] have to be discussed in order to implement smart home concepts broadly.

2.4   Platform-Concepts and IoT

In this day and age SHS often make use of the advantages created by progress in research fields which involve the internet and web-based products. Therefore, IoT [22] and web-based clouds [30] are integrated which leads to new platform concepts as well as innovative IoT-blockchains [14].

Platform Requirements and Possibilities. In general, platforms for SHS can be structured locally or globally. A local system with finite range has to be considered. Here, the gathering and evaluation of data stay in ones home which requires a sufficiently powerful system coping with all needed computations. Global systems mostly link data recording (local) with data analysis and storage (global) via the internet. Thereby, information gets transferred to an external server which enables the utilization of a less powerful system at home as well as data sharing and the usage of web applications. Nevertheless, local and global platform-concepts have to fulfill certain generic requirements, such as safety, protection of privacy, fully functioning communication as well as a clear, predefined recording of data. Global IoT platforms enable a focal view on ongoing processes as well as on the evaluation of captured data nonstop. Hence, such concepts provide a maximal complete description of the current state of the system at all time, which also supports central and instantaneous control and visualization possibilities. Due to these facts, a well designed platform is inevitable and therefore, one should anticipate a more detailed assessment of present platform-concepts and their architecture. An example for a generic IoT platform architecture is given in [24], which was also mentioned in Sect. 1.2.

Communication. A key feature of SHS is the communication between the different elements of the system. Sensors and actuators have to exchange information with the base of the system and, regarding IoT, this base has to connect with further web-based applications, such as clouds or external servers. For this reason, communication has to be stable, fast and secure. This can be achieved by utilizing proper communication protocols. [26], for example, presents an overview of wireless IoT protocols and discusses their properties in regards of security. Certainly, one has to decide on the type of connection (wired or wireless, one way or two way, etc.) beforehand. Furthermore, in comparison with the section above, local as well as global communication is feasible. Data transfer via a local internet-based wireless system [19], for example, facilitates a resilient link between various elements of a SHS.

Data Evaluation and Storage. In Sect. 2.1, the usability of SHS was discussed. In conjunction with IoT, this generic criterion depends on an appropriate evaluation of gathered information. Since data evaluation plays an important role, this topic should not be overlooked while assessing SHS. Thereby, a suitable choice of data for the analysis as well as matching methods are crucial. An example of a big data evaluation approach regarding energy management systems is given in [1]. For the purpose of storing user data, definite parameters have to be fastened. The leading questions are: How long should data be stored? What accuracy is necessary to reproduce records completely? Where should the storage be located? How save has the data depot to be in order to prevent leaks of private information?

2.5   Safety and Privacy

This section can be divided into three separate partitions coexisting in a ideally functioning SHS. Safety corresponds to circumstances which secure users as well as their surroundings; privacy depicts issues regarding the transfer and storage of sensible user-data; risk analysis uses surveillance algorithms to check the system on various malfunctions.

General Safety Issues. Issues concerning safety is covered in the ubiquitous research field of safety engineering [35]. Proper safety engineering assures that a system stays predictive and that its critical behavior produced by malfunctions can be analysed, foreseen and prevented. This also holds for SHS; systems have to be secure for their users and surroundings. For this purpose, possible electrotechnical problems and malfunctions have to be parsed beforehand and it is also necessary to propose safety protocols which control procedures ensuring a safe and secure usage. In [11] these topics are discussed with focus on children and elderly people as a part of the user-spectrum.

Privacy and Data Protection. SHS and SHA evaluate and, eventually, store sensitive user-data. Hence, insecure data protection yields threats regarding privacy. Such issues could enable a third person to intrude private space and to obtain sensitive data. This also affects, e.g., smart devices in the framework of SHS [20]. Due to the close connection with IoT and the concept of platforms, data protection has to be assessed in conjunction with web-based services and wireless communication. In order to enable a secure data transfer and storage, suitable protocols and procedures, which focus on efficiency and a very low failure rate, have to be included in SHS. However, most important is the user perception of personal space and data protection [38]. Users should be aware of the possible hazards regarding privacy in order to preserve personal information while using SHA.

Risk Analysis. Various risk detection algorithms ensure a fully working SHS. Such algorithms spot malfunctions and intruders immediately while reviewing internal processes of the system. In order to retain safety and privacy protection, proper measures have to be chosen and, therefore, included in an evaluation of SHS. In [6], for example, a recently developed defense mechanism, called Stochastic Traffic Padding, is introduced and reviewed.

Finally, it is to note that the above reviewed criteria, in conjunction with the decision table proposed in the following section, serve as a guideline to assess SHS. Hence, the given references in Sect. 2 should enable a discussion in greater detail.

3   Decision Making with Analytic Hierarchy Process

By incorporating the above listed criteria into the Analytic Hierarchy Process (AHP), see [31], a usable basis for assessing SHS and SHA is achieved. In this section the implementation of criteria into the AHP as well as advantages and disadvantages of the process regarding SHS are going to be discussed. The procedure itself will not be reviewed. For more detail on the method and calculations consider [15] or [34].

The AHP represents a structured, criteria based procedure in order to deal with complex choices resulting in a sophisticated and comprehensible decision. In general, the AHP requires at least two objects at choice, serving as the bottom of the hierarchy (see Fig. 1), which can be weighted by a set of independent criteria in order to achieve a satisfactory goal, which represents the top of the hierarchy in Fig. 1. In the case of SHS and SHA the in Sect. 2 given criteria provide such a suitable set and can therefore be used in order to apply the AHP, fulfilling the role of the second level in the hierarchy picture. These criteria have to be clearly separated to enable a utilization of the AHP. One example for a distinct

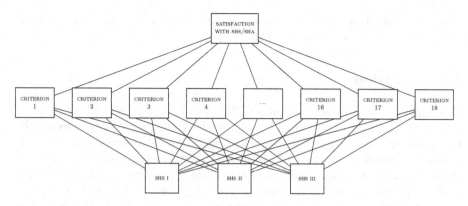

Fig. 1. The decomposition of the decision making problem regarding three different subjects into hierarchy provided by the AHP for SHS and SHA is presented.

disposition of criteria is shown below. It is to mention, that this list represents a very detailed segmentation, which is not always needed. Therefore, one could shorten this list by omitting particular issues depending on the satisfactory goal or even combine certain criteria. The numbering of this list is oriented towards the structure of Sect. 2 and further explanations to each criterion can also be found there.

1 General Features	10 Complexity
2 Controlling	11 Investment of Money
3 Visualization	12 Investment of Time
4 Added Values	13 Platform-concept (IoT)
5 Improvements/Upgrades	14 Communication
6 Energy/Ecological-awareness	15 Data Evaluation
7 Custom Features	16 Security
8 Additional Integration	17 Protection of Privacy
9 User-adaptation	18 Risk Analysis

The AHP results in an unambiguous statement about the preference after comparing different SHS or SHA. Furthermore, it also provides information on how close the different objects are to the ideal satisfactory goal. As a consequence, it gives an overall result weighting the different systems based on the applied criteria. On the one hand, the AHP is a broadly used and well established method and it distinguishes itself in a structured procedure and clear results. These features enable a relatively unbiased and subject unrelated decision, which is very important in the field of SHS due to the large span of different assessment topics. On the other hand, the AHP is a sophisticated tool which might not be utilized by users and therefore, it restricts its overall applicability. However, it provides a useful tool for developers, research groups as well as commercial businesses. All things considered, applying the AHP with the list of criteria given above yields a consistent assessment and decision entailing various fields of interest concerning SHS and SHA.

4   Conclusion

This paper reviews the most important criteria in order to assess and gauge smart home systems. It also provides a decision table applying these criteria explicitly. First, usability and sustainability are discussed. General features as well as controlling and visualization have to add values to the domestic or professional lives of users. In order to ensure a sustainable investment, updates and improvements have to be feasible while maintaining energy-awareness and ecological sensibility. By keeping the complexity low and enabling additional integration as well as custom features, a self-adjustable and continuous user-adaptation is provided. Accompanied by platform-concepts and IoT, smart home systems close up to the current century which results in new opportunities that include the utilization of cloud services and web-based applications. Lastly, safety issues, privacy protection and risk analysis should not be overlooked since these criteria yield a safe and secure usage of smart home applications. The in Sect. 3 given list of criteria facilitates a practical assessment using the Analytic Hierarchy Process. This procedure in conjunction with the presented criteria leads to commensurable results and a deliberate decision.

Acknowledgement. The K-Project Dependable, secure and time-aware sensor networks (DeSSnet) is funded within the context of COMET Competence Centers for Excellent Technologies by the Federal Ministry for Climate Action, Environment, Energy, Mobility, Innovation and Technology (BMK), the Federal Ministry for Digital and Economic Affairs (BMDW), and the federal states of Styria and Carinthia. The program is conducted by the Austrian Research Promotion Agency (FFG). The authors are grateful to the institutions funding the DeSSnet project and wish to thank all project partners for their contributions.

References

1. Al-Ali, A.R., Zualkernan, I.A., Rashid, M., Gupta, R., Alikarar, M.: A smart home energy management system using IoT and big data analytics approach. IEEE Trans. Consum. Electron. **63**(4), 426–434 (2017)
2. Alaa, M., Zaidan, A., Zaidan, B., Talal, M., Kiah, M.: A review of smart home applications based on internet of things. J. Netw. Comput. Appl. **97**, 48–65 (2017)
3. Alam, M.R., Reaz, M.B.I., Mohd Ali, M.: A review of smart homes - past, present, and future. IEEE Trans. Syst. Man Cybern. **42**, 1190–1203 (2012)
4. Anthi, E., Williams, L., Słowińska, M., Theodorakopoulos, G., Burnap, P.: A supervised intrusion detection system for smart home IoT devices. IEEE Internet Things J. **6**(5), 9042–9053 (2019)
5. Anvari-Moghaddam, A., Monsef, H., Rahimi-Kian, A.: Optimal smart home energy management considering energy saving and a comfortable lifestyle. IEEE Trans. Smart Grid **6**(1), 324–332 (2015)

6. Apthorpe, N., Huang, D., Reisman, D., Narayanan, A., Feamster, N.: Keeping the smart home private with smart(er) IoT traffic shaping. Proc. Privacy Enhancing Technol. **2019**(3), 128–148 (2019)

7. Beckel, C., Serfas, H., Zeeb, E., Moritz, G., Golatowski, F., Timmermann, D.: Requirements for smart home applications and realization with ws4d-pipesbox. In: ETFA 2011, pp. 1–8. IEEE (2011)

8. Brdiczka, O., Crowley, J.L., Reignier, P.: Learning situation models in a smart home. IEEE Trans. Syst. Man Cybern. **39**(1), 56–63 (2009)

9. Brush, A., Lee, B., Mahajan, R., Agarwal, S., Saroiu, S., Dixon, C.: Home automation in the wild: challenges and opportunities. In: CHI 2011 Proceedings of the SIGCHI Conference on Human Factors in Computing Systems, pp. 2115–2124. ACM Conference on Computer-Human Interaction (2011)

10. Chan, M., Campo, E., Estève, D., Fourniols, J.Y.: Smart homes - current features and future perspectives. Maturitas **64**(2), 90–97 (2009)

11. Chitnis, S., Deshpande, N., Shaligram, A.: An investigative study for smart home security: issues, challenges and countermeasures. Wirel. Sens. Netw. **8**, 77–84 (2016)

12. Demiris, G., Hensel, B.: Technologies for an aging society: a systematic review of "smart home" applications. Yearb. Med. Inform. **3**, 33–40 (2008)

13. Dewsbury, G., Clarke, K., Rouncefield, M., Sommerville, I., Taylor, B., Edge, M.: Designing acceptable 'smart' home technology to support people in the home. Technol. Disabil. **15**(3), 191–199 (2003)

14. Dorri, A., Kanhere, S.S., Jurdak, R., Gauravaram, P.: Blockchain for IoT security and privacy: the case study of a smart home. In: IEEE International Conference on Pervasive Computing and Communications Workshops, pp. 618–623. IEEE (2017)

15. Forman, E.H., Gass, S.I.: The analytic hierarchy process-an exposition. Oper. Res. **49**(4), 469–486 (2001)

16. Gross, C., Siepermann, M., Lackes, R.: The acceptance of smart home technology. In: Buchmann, R.A., Polini, A., Johansson, B., Karagiannis, D. (eds.) BIR 2020. LNBIP, vol. 398, pp. 3–18. Springer, Cham (2020). https://doi.org/10.1007/978-3-030-61140-8_1

17. Guinard, D., Trifa, V., Mattern, F., Wilde, E.: From the internet of things to the web of things: resource oriented architecture and best practices. In: Uckelmann, D., Harrison, M., Michahelles, F. (eds.) Architecting the Internet of Things, pp. 97–129. Springer, Heidelberg (2011). https://doi.org/10.1007/978-3-642-19157-2_5

18. Hargreaves, T., Wilson, C., Hauxwell-Baldwin, R.: Who uses smart home technologies? Representations of users by the smart home industry. In: European Council for an Energy Efficient Economy (ECEEE) Summer Study on Energy Efficiency in Buildings (2013)

19. Jiang, H., Cai, C., Ma, X., Yang, Y., Liu, J.: Smart home based on WiFi sensing: a survey. IEEE Access **6**, 13317–13325 (2018)

20. Kang, W.M., Moon, S.Y., Park, J.H.: An enhanced security framework for home appliances in smart home. HCIS **7**(1), 1–12 (2017)

21. Kim, J.E., Barth, T., Boulos, G., Yackovich, J., Beckel, C., Mosse, D.: Seamless integration of heterogeneous devices and access control in smart homes and its evaluation. Intell. Build. Int. **9**, 1–18 (2015)

22. Tan, L., Wang, N.: Future internet: the internet of things. In: 3rd International Conference on Advanced Computer Theory and Engineering (ICACTE), vol. 5, pp. 376–380. IEEE (2010)

23. Madakam, S., Ramaswamy, R.: Smart homes (conceptual views). In: 2014 2nd International Symposium on Computational and Business Intelligence, pp. 63–66. IEEE (2014)
24. Malche, T., Maheshwary, P.: Internet of things (IoT) for building smart home system. In: International Conference on I-SMAC (IoT in Social, Mobile, Analytics and Cloud) (I-SMAC), pp. 65–70. IEEE (2017)
25. Marikyan, D., Papagiannidis, S., Alamanos, E.: A systematic review of the smart home literature: a user perspective. Technol. Forecast. Soc. Chang. **138**, 139–154 (2019)
26. Marksteiner, S., Expósito Jiménez, V.J., Valiant, H., Zeiner, H.: An overview of wireless IoT protocol security in the smart home domain. In: Internet of Things Business Models, Users, and Networks, pp. 1–8. IEEE (2017)
27. Mehmood, F., Ullah, I., Ahmad, S., Kim, D.: Object detection mechanism based on deep learning algorithm using embedded IoT devices for smart home appliances control in cot. J. Ambient. Intell. Humaniz. Comput. **10**, 1–17 (2019)
28. Park, S.H., Won, S.H., Lee, J.B., Kim, S.W.: Smart home - digitally engineered domestic life. Pers. Ubiquit. Comput. **7**(3), 189–196 (2003)
29. Piyare, R., Lee, S.R.: Smart home-control and monitoring system using smart phone. ICCA ASTL **24**, 83–86 (2013)
30. Ray, P.P.: A survey of IoT cloud platforms. Future Comput. Inform. J. **1**(1), 35–46 (2016)
31. Saaty, T.L.: What is the analytic hierarchy process? In: Mitra, G., Greenberg, H.J., Lootsma, F.A., Rijkaert, M.J., Zimmermann, H.J. (eds.) Mathematical Models for Decision Support, pp. 109–121. Springer, Heidelberg (1988). https://doi.org/10.1007/978-3-642-83555-1_5
32. Solaimani, S., Keijzer-Broers, W., Bouwman, H.: What we do - and don't - know about the smart home: an analysis of the smart home literature. Indoor Built Environ. **24**(3), 370–383 (2015)
33. Vanus, J., Kucera, P., Koziorek, J., Machacek, Z., Martinek, R.: Development of a visualisation software, implemented with comfort smart home wireless control system. In: Park, J.J.J.H., Stojmenovic, I., Jeong, H.Y., Yi, G. (eds.) Computer Science and its Applications, pp. 581–589. Springer, Berlin Heidelberg (2015). https://doi.org/10.1007/978-3-662-45402-2_84
34. Vargas, L.G.: An overview of the analytic hierarchy process and its applications. Eur. J. Oper. Res. **48**(1), 2–8 (1990)
35. Verma, A.K., Ajit, S., Karanki, D.R., et al.: Reliability and Safety Engineering, vol. 43. Springer, London (2010). https://doi.org/10.1007/978-1-84996-232-2
36. Yang, H., Lee, W., Lee, H.: IoT smart home adoption: the importance of proper level automation. J. Sens. **2018**, 1–11 (2018)
37. Yang, R., Newman, M.W., Forlizzi, J.: Making sustainability sustainable: challenges in the design of eco-interaction technologies. In: Proceedings of the SIGCHI Conference on Human Factors in Computing Systems, pp. 823–832. Association for Computing Machinery (2014)
38. Zheng, S., Apthorpe, N., Chetty, M., Feamster, N.: User perceptions of smart home IoT privacy. Proc. ACM Hum.-Comput. Interact. **2**, 1–20 (2018)

Advances in Decision Support Systems' Technologies and Methods

Modelling the Effects of Lockdown and Social Distancing in the Management of the Global Coronavirus Crisis - Why the UK Tier System Failed

Jinyi Liu[1] and Patrick Stacey[2]([✉])

[1] Loughborough University, Loughborough, UK
[2] Centre for Information Management, Loughborough University, Loughborough, UK
p.stacey@lboro.ac.uk

Abstract. This article reviewed the current situation of coronavirus crisis, as well as policies regarding disease control implemented by China, Italy, UK and U.S. We analysed the coronavirus development trend by visualizing data with Python. We analysed the impact of social distancing and performed simulations of a variety of lockdown circumstances using NetLogo. Throughout our analysis, we examined the importance of social distancing, lockdown and quarantine measures, and extended insights on the ineffectiveness of the UK Covid-19 Tier system.

Keywords: Coronavirus · Simulation · Pandemic control · Social studies · Policymaking

1 Introduction and Literature Review

The purpose of this paper is to study and model the spread of Covid-19, particularly *vis a vis* various states of lockdown. Our research question is: Why did the UK Tier system fail to curb transmission rates? No other study has modelled this. We provide simulations which contrast the dynamics and outcomes of three different scenarios: no lockdown, partial lockdown/Tier system, and complete lockdown. We contextualize this with trends from four different countries. The scientific and obvious conclusion is unassailable - only complete lockdowns are effective in curbing transmission. However, what is surprising is that we predicted and modeled that a partial lockdown would only serve to exacerbate transmission compared to a no lockdown scenario. We encourage policymakers and practitioners to reconsider partial lockdown strategies and use our models and reasoning to understand why they are ineffective.

The coronavirus disease 2019 is caused by the severe SARS-CoV-2 virus (Mayo Clinic Staff 2020), resulting in, at the time of writing, over 48 million cases and 1.2 million deaths worldwide[1]. According to the U.S. CDC, the transmission of the virus occurs when contracted patient's release droplets from their mouth or nose into the air

---

[1] https://coronavirus.jhu.edu.

U. Jayawickrama et al. (Eds.): ICDSST 2021, LNBIP 414, pp. 69–83, 2021.
https://doi.org/10.1007/978-3-030-73976-8_6

by either coughing, sneezing, or talking (Centres for Disease Control and Prevention 2020). The virus can endure on various types of surfaces for up to 28 days (Riddell et al. 2020). Much of this is common knowledge now of course. What is less known in lay terms and many other domains, except the medical profession, is that established means of testing is not the most effective method, For example, it is not commonly known that sputum samples are more reliable as a form of detection; this is because the viral load is higher than in Nasopharyngeal Airways (NPAs) and endure longer (Liu et al. 2020). This calls into question current approaches to testing in the UK and elsewhere and could be a factor behind the volatile success rate in detection. Secondly, at the population level, researchers predict that 2% of asymptomatic coronavirus carriers are responsible for 5%–10% of acute respiratory infections (Cascella et al. 2020). There is a cultural assumption that people who look well, are well. Understanding this and the 'deceitful' nature of the coronavirus is key to future prevention.

Another facet is that people trust others within their social networks - they assume their friends and such like are trustworthy in terms of their apparent health. This can be fallacious. Consider recent work on social connectiveness which models the connection between two or more people nodes - a study used Facebook users' friendship networks to understand how the coronavirus was spreading (Kuchler et al. 2020). The authors used aggregated data from Facebook to show that COVID-19 was more likely to spread between areas with stronger social network connections. They found that areas with more social ties generally had more confirmed COVID-19 cases (Ibid:p.1), for example Westchester County, NY, in the U.S. and Lodi province in Italy. Further to Ibid, geographical tracking and mapping technology also helps in identifying regional pandemic situations. Boulos and Geraghty (2020) provided various types of coronavirus tracking tools in their paper, including: the JHU Center for Systems Science and Engineering dashboard, the WHO dashboard, HealthMap, Chinese coronavirus 'close contact detector' geosocial app, WorldPop and EpiRisk. We are minded by these tools to take a comprehensive perspective on modeling and understanding the spread of the pandemic. We now review some specific approaches to analyzing the transmission of coronavirus.

Dey et al. (2020) agree that by monitoring the coronavirus outbreak using highly available computing tools such as 'Python' one can gather and disseminate detailed figures and trends of different location's situation (e.g. Kershaw et al. 2014), providing significant evidence for scientific communities to make independent assessments. Indeed, Fang et al. (2020) performed a simulation of the transmission dynamics of Covid-19 and its impact using Python. Part and parcel of their study was a sensitivity analysis to plot the R number, i.e. the Reproduction number; the R number represents how many people will be infected by one carrier. Their model (Ibid) possessed a coefficient of determination approaching 1 with a fitting bias of 3.02%; their model was therefore statistically highly precise.

This was made possible through the use of Python libraries such as Pandas. As a result, they concluded that the R number was increasing until authorities' intervention reversed the trend. Further, Python possesses real-time estimation capability, which helps trace real-time infection mortality and risk assessment. Jung et al. (2020) modelled epidemic growth from two perspectives: a single index case (patient zero) with illness onset and the growth rate fitted along with related parameters and exported cases. Their research predicted the crude case-fatality risk (cCFR) to be in a value between 5% to 8%, with an R_0 between 1.6 to 4.2; R_0 is the basic reproduction number which differs to the oft-touted R number in the media, which is called the *effective* reproduction number. The former is a raw number whereas the latter imputes an average likelihood of transmission and is therefore considered more real-world since not everyone succumbs to said transmitted disease. Using Python in combination with other tools such as *Stata* enables one to also study the economic impacts of transmission dynamics. For example, Fernandes (2020) examined the supply chains and stock market figures respectively, making comparisons with historical data, and hypothesized at a global level that every additional month of pandemic yielded crisis costs equivalent to 3% of global GDP.

In summary, the research on the coronavirus outbreak has yielded an effective tool-set for analyzing the pandemical spread in terms of locales, R numbers and economic impacts. A great deal of innovative research has been done that has improved understanding over how the coronavirus has spread, as well as how it can be monitored. We found that Python is a popular tool therein, combined with other GIS transformation techniques and statistical packages, to explore the spreading mechanism and geographical differences of Covid-19. Furthermore, the literature suggested to us that a detailed study was required in terms of social, structural phenomena such as social distancing, lockdown and medical capacity. Our contributions to knowledge are: (i) a comprehensive modelling approach to understanding spread and assessing strategies of containment; (ii) philosophical reasoning drawing on Bourdieu (1984) to explain why a partial lockdown worsened transmission compared to a no lockdown scenario.

2 Method

In this section we cover data collection and analysis approaches. We collected and ingested datasets from Johns Hopkins University's real-time coronavirus tracing system, and Oxford University's project in exploring the testing sample number, with the aim of ensuring the authenticity, accuracy and completeness of data. The JHU's dataset contains confirmed cases in each country, mortality numbers, recovery numbers, with some exponential curves' analyses showing the developing trend in locations like Hubei Province in China, Italy and the United States. However, some countries do not report complete figures; for example, the UK does not report recovery cases and Korea does not report death cases. This limited our analysis and limits the modelling that public health experts can perform per se. This is a moot example of how politics can affect data science and public health preparations. To assist us in making comparisons between different countries' pandemic medical capability we draw on the coronavirus testing dataset published by *Our World in Data* founded by Oxford University.

This reports daily Covid-19 testing numbers for a variety of countries. We selected four major countries' Covid-19 cases for study and produced visualizations, drawing on the literature review for constant cross-checking and comparisons. Table 1 reports on the data we collected/ingested:

Table 1. Summary of data collection

Dataset	No. rows	No. cols	Total cells
Confirmed cases	268	292	78256
Death cases	268	292	78256
Recovered cases	255	292	74469
Meta data cases	190	14	2660
Total cells of data			*233641*

The rows consisted of countries while the columns contained the respective frequencies of cases, deaths etc. from 22nd January 2020 until 3rd November 2020 (286 days). This, therefore, from a UK/European perspective includes data before and after the lockdowns of the first pandemical wave, as well as data into the partial lockdowns of the second wave. We used Python to construct ranking tables and graphs, detailed infection cases number, recovery and death rate number, and developing trends in an exponential curves format. We used the following python libraries for these analyses: *pandas, numpy, plotly.graph_objects, plotly.express, matplotlib.pyplotimport* and *NetLogo*. Based on insights from the literature, we used NetLogo, a pythonic Agent-Based Model (ABS) simulation software, to simulate the spread of the pandemic. Doing so showed geographical differences in disease spreading. Building such a simulation requires knowledge of virology to set up coefficients and indexes. For this, we used a pre-built multi-agent model that factored in population, clinical parameters, and the following social connection factors (Ronald 2020): primary population per age group, infection possibility per age group, mortality possibility per age group, commuting population per day and infection rate in commuters. For the prediction, the Susceptible, Infectious, or Recovered (SIR) model was used to predict future trends and lockdown measure effectiveness. What is more, the Susceptible, Exposed, Infectious, or Recovered (SEIR) model was built to assess the effects of social distancing and isolation. The medical supply capability, primary patients' number, average infection length, average time before diagnosed, average days for spreading, maximum distance for spreading, severe rate, and death reason in terms of a severe infection were considered because they helped visualizing the enlarging pandemic spread, as well as studying prevention. Further factors such as schools and colleges numbers, hospitality capacity or amount, and public places open number were added to study if and how social distancing effects the spread of the disease.

3 Trend Analysis

The analyses in this section include coronavirus case trends for China, Italy, UK and the US. Firstly, from our dataset of approximately ¼ million datapoints, representing 286 days of data (22nd January 2020 until 3rd November 2020, we derived the following snapshot:

Table 2. Summary of Global Cases (286 days since 22nd January 2020)

Confirmed	Recovered	Deaths	Active
48,136,225	31,889,030	1,225,913	14,990,278

Table 2 shows that the total confirmed cases of Covid-19 exceeds 48 million at the time of writing, with a general recovery rate of 66.24% and a general death rate of 2.54% among diagnosed patients. As noted in the methodology, not all countries report death and recovery rates. From the trend data, we found a sharp increase in total confirmed cases since 16th March 2020. The average rate of change across the globe for cases, deaths and recovery are depicted in the Table below:

Table 3. The average rates of change globally

Avg % Change	Python Pandas command	Result
Cases	confirmed_ts_summary.pct_change().mean()	0.043
Recovered	recovered_ts_summary.pct_change().mean()	0.052
Deaths	death_ts_summary.pct_change().mean()	0.042

The above Table 3 shows that the average recovery growth rate is higher than the death rate, which is encouraging. Enhanced testing capability and actual testing performed has likely contributed to these statistics. We combined the exponential curves for the four countries in terms of confirmed, death, recovered and active cases. These countries' trends were chosen since they received the most coverage in the media (caveats regarding the reporting of figures applies). We present four charts below. From left to right and top to down we present China, Italy, UK and the US (Fig. 1):

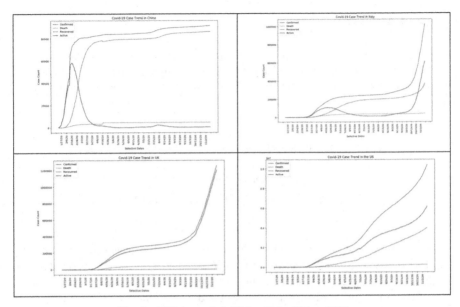

Fig. 1. CDRA trends in China, Italy, UK and US

3.1 Summary

As the first country attacked by the coronavirus, the recovery cases figure for China began to increase soon after confirmed cases rose. The graph above vividly shows us that within two to three months, the active cases declined sharply, while the recovered cases exceeded the active cases figure. The deaths cases curve slightly increased at first, then soon flattened. Draconian measures taken by China resulted in substantial achievements; it is a candidate exemplar country in handling Coronavirus per se. Before the Spring Festival, the city of Wuhan and the entire Hubei Province region went into lockdown (UKRI 2020). The infection rate reduced immediately after the travel ban was implemented. Furthermore, people were mandated to wear face-masks in many public areas, while outdoor activities were restricted by regulators (Cohen 2020). For those who entered large indoor spaces, their physical temperatures were tested, and any confirmed cases quarantined either in a hospital or mobile cabin hospital. Despite some political controversies over these harsh measures, China's effort in controlling the Covid-19 spread should be considered as one of the most successful and efficient ones. In addition, news suggesting further outbreaks of coronavirus in Jilin Province (Stanway et al. 2020), Beijing (BBC 2020a, b) and other places (TIME Magazine 2020) also tells us the consequences we may confront once we lower our guard before the Covid-19 is under control. As well as the success in China we also note the effectiveness of some of the Italian approaches (Mccall 2020).

We now discuss some of the points of inflection in the above graphs: for UK and Italy, the first wave took place in early March 2020. Over the Summer period, UK active cases continued to grow, albeit steadily, whereas Italian active cases declined for the same period. The earliest measure taken in Italy was January 31st 2020, with the mobilization of police and armed forces to the risky of Lombardy and Veneto (BBC 2020a, b). Also, local authorities promoted large scale testing, even retesting among residences, symptomatic or not (McCall 2020). Various isolation and medical tactics both led to a halt in the outbreak. The UK undertook one of the highest number of tests in Europe, just behind Italy, Spain and Germany (UK Department of Health Social Care 2020). And yet cases continued to rise in the UK under a virtual 'no lockdown' scenario. We surmise that UK's fortune of being an 'international transport hub' is also a curse - UK is a popular destination and hub which has contributed to the continuing spread of the disease steadily throughout the Summer and then more pronounced thereafter; our upcoming simulation models further enlighten this point. U.S. cases have also been steadily increasing since the outbreak in mid-March 2020. The inflection point of July 5th 2020 marks an increase in confirmed and active cases there. According to external news sources and our testing sample dataset, the U.S. testing number is the largest among other four countries we selected since March, which strongly suggested a reason why their confirmed cases become the highest; the more tested, the more discovered. In addition, to figure out why the total recovery rate (17.66%) in U.S. is lower than many other European countries like Italy (48.01%), we can consider some recent heated discussions over shortages of medical supply and medical overload (Schlanger 2020); the lack of protective and treatment equipment poses a huge threat on both patients and healthcare workers.

4 Simulation Model Analysis

Going beyond the above reporting of figures and trends, we conducted SEIR modelling for a variety of social distancing and lockdown scenarios using Python packages Scipy, Pandas and Matplotlib, as well as NetLogo. The SEIR model is useful in analyzing pandemic situations where incubation period applies to large amounts of individual cases. By importing the SEIRClass package, we simulated the process of disease spreading following the process: susceptible, exposed, infectious and recovered. The mechanism represented in this model can be explained as: after a pandemic outbreak, the population in contact with the virus will move from one of these four states to another. When an R-state is reached, infection will no longer proceed, which can be considered as an 'immune' state (Hubbs 2020). The SEIR model we used can be expressed mathematically as follows:

$$\Delta S = -\beta SI \tag{1}$$

$$\Delta E = \beta SI - \alpha E \tag{2}$$

$$\Delta I = \alpha E - \gamma I \tag{3}$$

$$\Delta R = \gamma I \tag{4}$$

$$N = S + E + I + R \tag{5}$$

In all these five equations, parameters α represents the inverse of the incubation period, that is $1/t\_incubation$. β refers to the average contact rate in the population. γ symbolizes the inverse of the infectious period mean value, known as $1/t\_infectious$. Equation (1) computes variations in susceptible cases - this is moderated by confirmed cases and contact between susceptible and confirmed. The second equation computes variations in exposed cases - increases per contact rate, and decreases with the incubation period. Equation (3) computes variations in infected cases resulting from exposed population and incubation length. It decreases with infectious period - a larger γ value yields quicker recovery which leads us to the final stage in (4). Equation (5) serves as a constraint, indicating zero existence of birth/migration effects in the model; the population is fixed throughout. R, which also known as R_0, refers to the speed of virus spreading and can be calculated as:

$$R_0 = \frac{\beta}{\gamma} = average\ contact\ rate \times infection\ period\ mean\ value.$$

Further, to model the social distancing effect, we introduced a new coefficient ρ, which is a constant term between 0 and 1. A $\rho = 0$ means a complete lockdown when everyone is isolated, and a $\rho = 1$ represents the same situation as the above math model. Hence, Eqs. (1) and (2) can be changed to the following form:

$$\Delta S = -\rho\beta SI \tag{1'}$$

$$\Delta E = \rho\beta SI - \alpha E \tag{2'}$$

4.1 Simulation Set 1 - Loughborough Campus

The first simulation we present is experimental, however, we had absorbed a lot of the literature and derivative norms from scientific studies of Covid-19; therefore, our assumptions are influenced by them to some degree. This simulation is based on the population of the Loughborough University Campus where N = 14,000. The parameters were referenced from medical research.. Referring to Table 4 below: (i) the R_0 and incubation period value drew on Hellewell et al. (Hellewell et al. 2020), (ii) the infection period drew on Peng et al. (2020), (iii) The social distancing value of 2 m recommended by many experts and medical organizations.

In the experiment in the upper chart of Figure 2 above, we assumed 40% effective social distancing. The result shows that at first the susceptible cases figure is very high and infectious cases close to zero. Then after seven to eight weeks, the susceptible cases decline by more than half, the infectious cases increase, and the recovered cases also largely increase by 78.57%. In the second experiment in the lower chart in Figure 2, we made a comparison between a no social distancing scenario and a 40% effective social distancing scenario. As we can see, without social distancing condition much of the population is exposed, the infection rate skyrockets; it only takes 4 to 5 weeks to reach more than 1000 infections. In comparison, with 40% effective social distancing, the exposed and infected cases trend curve flattens, and at around 6 to 7 weeks both trends peak at only 800 units. What we can further see is that under social distancing measures,

Table 4. Pythonic simulation - Loughborough university campus

Parameter	Value	Description
R0	3.5	Basic reproduction number
t_incubation	5.1	Incubation rate
t_infective	2	Infection rate
N	14000	Campus population
t_social_distancing	2	Metres

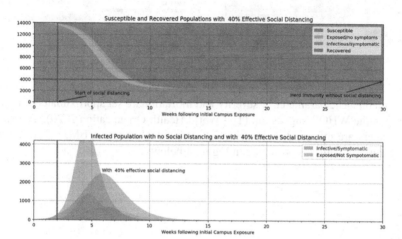

Fig. 2. Experimentally simulating SEIR rates at Loughborough University

recovered population is much larger when herd immunity is reached, comparing with no social distancing. This provides a convincing argument that social distancing is an effective weapon in combating Covid-19.

4.2 Simulation Set 2 - Agent Based Modelling

Having observed the above results from a relatively simple model, we decided to produce more simulations using more complex sets of parameters and more scenarios in order to investigate further. For the first in this second set of simulations, we set it up as a worst-case scenario by defining a variety of demographic, virological and medical service parameters; all done using NetLogo's Dashboard (Romero et al. 2020) - see Figure 3 below:

In this experiment, the majority of the population are in the 20–50s year old band, with a large aging group, and old people vulnerable to the coronavirus. Under this worstcase scenario no prevention measures are taken, and all public places are open. What's more, people take no social distancing measures. We assume the medical care capability is

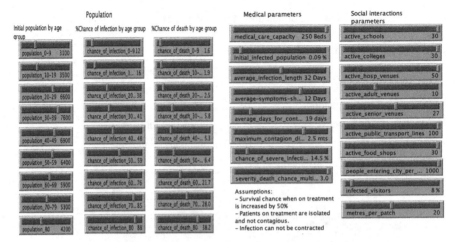

Fig. 3. NetLogo dashboard for simulation 2.1 - Worst case scenario

fully available. The parameter of severity and mortality are set approximately in accordance with the WHO's suggestion (The World Health Organization 2020). The values of parameters are informed by a variety of studies including the BMJ[2]. The result of running this simulation is presented in Figure 4 below:

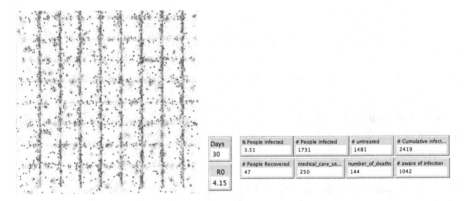

Fig. 4. Simulation 2.1 - Worst case scenario

Figure 4 show that a large number of people will be infected when no measures are taken. It suggests that 3.51% of the total population will be infected under so-called 'herd immunity'. The spatial area simulated in the left of the Figure is populated with red spots which represent patients spatially. Also, even with fully available medical care, nearly 85% of patients do not receive treatment on time, and 39.8% of patients are asymptomatically infected.

---

[2] https://www.bmj.com/content/369/bmj.m1327.

For the second simulation in this set, we set it up similar to the Tier system approach implemented in the UK - i.e. a partial lockdown. The results are presented in Figure 5:

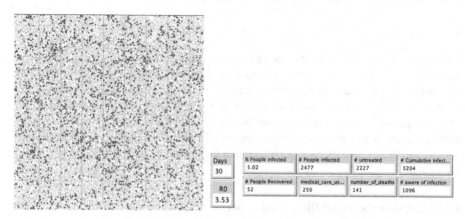

Days	% People infected	# People infected	# untreated	# Cumulative infect...
30	5.02	2477	2227	3204

	# People Recovered	medical_care_us...	number_of_deaths	# aware of infection
R0	52	250	141	1096
3.53				

Fig. 5. Simulation 2.2 - Partial Lockdown (UK Tier System)

This result surprised us; it presents an even worse situation than under no lockdown or social distancing. The percentage of infections rises and the spatial area in the left of the Figure is populated with more red spots which represent patients spatially. The infection rate dramatically rose to 5.02% within only 30 days. The only slight positive was that the death rate declined by three cases, and the recovery rate slightly rose.

For the third simulation in this set, we set it up as a total lockdown. We defined it so in the NetLogo dashboard, setting the social distancing as 4 metres this time. The results from this simulation are presented in Figure 6 below:

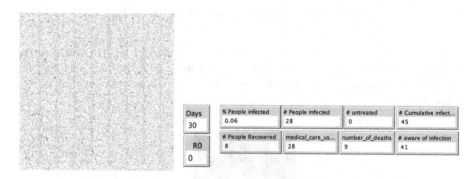

Days	% People infected	# People infected	# untreated	# Cumulative infect...
30	0.06	28	0	45

	# People Recovered	medical_care_us...	number_of_deaths	# aware of infection
R0	8	28	9	41
0				

Fig. 6. Simulation 3.2 - Total lockdown - Best case scenario

The effectiveness of this configuration given the sparsity of the red dots (patients) is due to a complete strict lockdown and maximum social distancing hereby shown. As a result, during a 30-day experimental period, only 0.06% of the population contracted the coronavirus, and a very small number of infection cases are detected. The general mass health situation remains at a relative smooth level, infection rate declines and the mortality exponential curve is flattened. While this result is not surprising, it provides an important computational benchmark as we continue with our analysis.

4.3 Summary

The above experiments make it clear that only a complete lockdown and greater social distancing is effective in significantly reducing the virus spread and mortality rate, as well as relieving pressure on the national health service; the infection rate is only 0.06%, compared to 5.02% under partial lockdown, and 3.51% under no-lockdown. The Tier system or partial lockdown actually contributes to the spread of the virus; this is counterintuitive but is consistent with what happened in the UK. It is interesting to note that we initially ran these simulations in early Summer 2020. This requires further analysis and discussion:

4.4 Analysing the Social Distancing Aspect

We can extract the social distancing problem as a single research object, by using another ABM model to test its effectiveness. The R_0 rate is simulated in accordance with previous literature cited (Fig. 7):

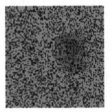

Fig. 7. Social distancing problem as a single research object

We set a lower social distancing value and observed the blue and red areas - these symbolize the alarming infection spread which takes up space in the total green region. However, by enlarging the social distancing value, the infected rate declines as per Figure 8 below:

Fig. 8. Social distancing problem as a single research object

5 Discussion and Implications

Some of our results are not surprising and reinforce the findings of other SEIR studies (e.g. Cascella et al. 2020), i.e. transmission is fast without a complete lockdown, and, only extreme social distancing measures help curtail the SEIR process. What *is* surprising and requires further discussion is why the SEIR process model is 'worse' under 'partial lockdown' compared to 'no lockdown'. We present a set of possible reasons below, while realizing the answer is complex: (i) social routines are potent habits and people have difficulty letting go of them - a great deal of sociological theory has shown this (e.g. Giddens 1984). This involves habitus (Bourdieu 1984) - a physical loci that enables and constrains routine social activity. Unless these loci are shutdown, pubs for example, there is little impact on the deep-seated habits of individuals and people will continue to seek out socialization in them. Under partial lockdown in the UK, depending on the Tier, many 'venues' for socialization remained open and therefore routine activity was promoted, (ii) the Tier system may have been confusing for much of the population too - only a national lockdown provides definitive constraints over routine social activity; (iii) Since it was not a national lockdown, the partial lockdown system may have given people a false sense of security; the situation was not perceived to be serious. However, these points do not explain why transmission was worse under partial lockdown compared to no lockdown. We suggest: (a) the anticipation of a possible Winter national lockdown motivated the UK population to ramp-up routine activity and make full use of the availability of socio-geographic resources, (b) a 'relief period' during partial lockdown in which people were still making up for lost or sacrificed routine activity during the first national lockdown, (c) being online more during the first lockdown could have led to individuals forging new contacts and enlarged social networks. This could have resulted in increased social activity during no/partial lockdown conditions; the relationship between trust and social networks has been researched before (Kuchler et al. 2020).

Going forward, populations will need to exercise more restraint even when the physical constraints on habitus have been relaxed, as well as exercise more autonomous reflexivity (Archer 2007) rather than adhering to social codes of contact within their social networks. Such reflexivity and restraint over social routines is a matter of survival in a pandemical scenario.

6 Conclusion

This report provides data, graphs and simulation experiments as evidence to support the thesis that Covid-19 is a highly infectious pandemic which has a strong spreading capability. The results reported show that the situation is still serious for every country and nation. We modelled how the UK Tier system (partial lockdown) worsened transmission rates. We offered a variety of reasons including routine social activity via the concepts of habitus, social networking and reflexivity.

References

Archer, M.S.: Making Our Way Through the World. Cambridge University Press, Cambridge (2007)

BBC: Coronavirus: Venice Carnival Closes as Italy Imposes Lockdown (2020a)

BBC: Coronavirus: Fear of second wave in Beijing after market outbreak (2020b)

Boulos, M.N.K., Geraghty, E.M.: Geographical tracking and mapping of coronavirus disease COVID-19/severe acute respiratory syndrome coronavirus 2 (SARS-Cov-2) epidemic and associated events around the world: how 21st century GIS technologies are supporting the global fight against Outbr. Int. J. Health Geogr. **19**(8), 12 (2020)

Bourdieu, P.: Distinction: A Social Critique of the Judgement of Taste. Trans. Richard Nice. Harvard University Press, Cambridge (1984)

Cascella, M., et al.: Features, Evaluation and Treatment Coronavirus (COVID-19) (2020)

Centres for Disease Control and Prevention (CDC): Coronavirus Disease 2019 (COVID-19)-Prevent Getting Sick (2020)

Chaudhary, J.: Data Modelling & Analysing Coronavirus (COVID19) Spread Using Data Science & Data Analytics in Python Code (2020)

Cohen, J.: Not Wearing Masks to Protect Against Coronavirus Is A 'Big Mistake,' Top Chinese Scientist Says (2020)

Dey, S.K., Rahman, M.M., Siddiqi, U.R., Howlader, A.: Analyzing the epidemiological outbreak of COVID-19: a visual exploratory data analysis approach. J. Med. Virol. **92**, 632–638 (2020)

Duddu, P.: Coronavirus in South Korea: COVID-19 Outbreak, Measures and Impact (2020)

Fang, Y., Nie, Y., Penny, M.: Transmission dynamics of the COVID-19 outbreak and effectiveness of government interventions: a data-driven analysis. J. Med. Virol. **92**, 645–659 (2020)

Fernandes, N.: Economic Effects of Coronavirus Outbreak (COVID-19) On the World Economy (2020)

Hayes, R.W.: Coronavirus: Japan's Low Testing Rate Raises Questions (2020)

Johns Hopkins University: COVID-19 Dashboard by The Center for Systems Science and Engineering (CSSE) At Johns Hopkins University (JHU) (2020)

Hellewell, J., et al.: Feasibility of controlling COVID-19 outbreaks by isolation of cases and contacts. Lancet Glob Health **8**, e488–e496 (2020)

Hubbs, C.: Social Distancing to Slow the Coronavirus, 12 March 2020. Towards Data Science. https://towardsdatascience.com/social-distancing-to-slow-the-coronavirus-768292f04296. Accessed 25 Jan 2021

Jung, S.-M., et al.: Real-time estimation of the risk of death from novel coronavirus (COVID-19) infection: inference using exported cases. J. Clin. Med. **9**(2), 523–533 (2020)

Kershaw, D. Rowe, M., Stacey, P.: Towards tracking and analysing regional alcohol consumption patterns in the UK through the use of social media. In: Proceedings of the 2014 ACM Conference on Web Science, June 2014, pp. 220–228 (2014). https://doi.org/10.1145/2615569.2615678

Kuchler, T., Russel, D., Stroebel, J.: The Geographic Spread of Covid-19 Correlates With Structure of Social Networks As Measured By Facebook, 1050 Massachusetts Avenue (2020)

Mayo Clinic Staff: Coronavirus Disease 2019 (COVID-19) Symptoms and Causes (2020). https://www.mayoclinic.org/diseases-conditions/coronavirus/symptoms-causes/syc-20479963

Mccall, R.: Coronavirus Mass Testing Experiment in Italian Town Appears to Have Halted Covid-19 Outbreak (2020). https://www.newsweek.com/coronavirus-mass-testing-experiment-italian-town-covid-19-outbreak-1493183

Mccurry, J.: Japanese Flu Drug 'Clearly Effective' in Treating Coronavirus, Says China (2020). https://www.theguardian.com/world/2020/mar/18/japanese-flu-drug-clearly-effective-in-treating-coronavirus-says-china

Niesen, R.K., Fletcher, R., Newman, N., Brennen, J.S., Howard, P.N.: Navigating the 'Infodemic': How People in Six Countries Access and Rate News and Information About Coronavirus. Reuters Institute, and Oxford University (2020). https://reutersinstitute.politics.ox.ac.uk/infodemic-how-people-six-countries-access-and-rate-news-and-information-about-coronavirus

Osaki, T.: How far can Japan Go to Curb the Coronavirus Outbreak? Not as far as You May Think (2020). https://www.japantimes.co.jp/news/2020/03/01/national/japan-coronavirus-outbreak/

Peng, L., et al.: Epidemic analysis of COVID-19 in China by dynamical modeling, BMJ Yale (2020). https://www.medrxiv.org/content/https://doi.org/10.1101/2020.02.16.20023465v1

Riddell, S., Goldie, S., Hill, A., et al.: The effect of temperature on persistence of SARS-Cov-2 on common surfaces. Virol. J. **17**, 145 (2020)

Romero, C.J., Tisnes, A., Linares, S.: Covid 19 Contagion Dynamics (2020). https://modelingc ommons.org/

Ronald, P.N.: Covid-19, coronary virus, control of virus infection (2020). https://modelingcomm ons.org/tags/one_tag/2100

Schlanger, Z.: Begging for Thermometers, Body Bags, and Gowns: U.S. Health Care Workers are Dangerously Ill-Equipped to Fight COVID-19 (2020)

Stanway, D., Lee, S.Y., Zhang, L.: China's Jilin City Imposes Travel Restrictions After New Coronavirus Cases (2020). https://www.reuters.com/article/us-health-coronavirus-china/chinas-jilin-city-imposes-travel-restrictions-after-new-coronavirus-cases-idUSKBN22P0FE

The World Health Organization: Q&A on Coronaviruses (COVID-19) (2020)

Tian, H., et al.: The Impact of Transmission Control Measures During the First 50 Days of the COVID-19 2 Epidemic in China (2020)

TIME Magazine: China Appears to Have Tamed a Second Wave of Coronavirus in Just 21 Days with No Deaths (2020)

UK Department of Health & Social Care: Coronavirus (COVID-19): Scaling Up Our Testing Programmes (2020)

UKRI: What Can We Learn About COVID-19 Control from China and Other East Asian Countries? (2020)

Woo, P.C., et al.: Characterization and complete genome sequence of a novel coronavirus, coronavirus HKU1, from patients with pneumonia. J. Virol. **79**(2), 884–895 (2005)

A Case Study Initiating Discrete Event Simulation as a Tool for Decision Making in I4.0 Manufacturing

Kristina Eriksson[1]([⊠]) [iD] and Ted Hendberg[2]

[1] University West, 461 86 Trollhättan, Sweden
kristina.eriksson@hv.se
[2] Siemens Energy AB, Kardanvägen 4, 461 38 Trollhättan, Sweden

Abstract. Smart manufacturing needs to handle increased uncertainty by becoming more responsive and more flexible to reconfigure. Advances in technology within industry 4.0 can provide acquisition of large amounts of data, to support decision making in manufacturing. Those possibilities have brought anew attention to the applicability of discrete event simulation for production flow modelling when moving towards design of logistics systems 4.0. This paper reports a study investigating challenges and opportunities for initiation of discrete event simulation, as a tool for decision making in the era of industry 4.0 manufacturing. The research has been approached through action research in combination with a real case study at a manufacturing company in the energy sector. The Covid-19 pandemic fated that adjusted and new ways of communication, collaboration, and data collection, in relation to the methods, had to be explored and tried. Throughout the study, production data, such as processing times, have been collected and analyzed for discrete event simulation modelling. The complexity of introducing discrete event simulation as a new tool for decision making is highlighted, where we emphasize the human knowledge and involvement yet necessary to understand and to draw conclusions from the data. The results also demonstrate that the data analysis has given valuable insights into production characteristics, that need addressing. Thus, revealing opportunities for how the initiative of introducing discrete event simulation as an anew tool in the wake of industry 4.0, can act as a catalyst for improved decision making in future manufacturing.

Keywords: Discrete Event Simulation · Decision support systems in manufacturing · Industry 4.0

1 Introduction

In the current era of Industry 4.0 (I4.0), the manufacturing industry is forming strategies for the application of new technologies towards increased digitization [1, 2]. There are several key technologies of I4.0 with relevance for the manufacturing sector i.e., the industrial Internet of Things (IoT), Cloud Computing, Big Data, Simulation, Augmented Reality, Additive Manufacturing, Horizontal and Vertical Systems Integration,

U. Jayawickrama et al. (Eds.): ICDSST 2021, LNBIP 414, pp. 84–96, 2021.
https://doi.org/10.1007/978-3-030-73976-8_7

Autonomous Robots and Cyber Security [3]. Among technological drivers for I4.0 are modeling and simulation-based development of productions systems [4], which can further their pertinency through technologies for real-time data collection and analysis to provide information to the manufacturing system [1, 5]. A field of knowledge that can contribute as a technological driver for I4.0 is Discrete Event Simulation (DES) [6]. In manufacturing settings DES has traditionally been applied to certain scenarios, such as identifying bottlenecks, experimenting with new factory layouts, and studying variations in the production flora [7, 8]. However, there is a potential for extended applications of DES as a tool for decision making in the era of I4.0. For example, automatic model generation and data exchange between manufacturing applications [9], coupling of modeling of production flows with multiagent-based simulation [10] and multi-objective optimization [11]. The advancement of real-time data acquisition has improved possibilities of such inputs into DES models [12]. Manufacturing today needs to handle heightened variability, uncertainty, and randomness, meaning smart factories need to become more responsive and faster to reconfigure [13]. Advancement of technologies can affect the application of production flow simulations, for example, easier and faster acquisition of large amounts of data for analysis to support decision making to deal with manufacturing variability and uncertainty. Hence, the possibilities of technologies within I4.0 have brought anew attention to the applicability of DES when moving towards design of logistics systems 4.0 [14]. Yet there exist barriers towards successful I4.0 implementation such as, insufficient digital skills and resources, ineffective change management, and lack of digital strategies [15]. There are theoretical discussions on how to move forward i.e. [16] suggest a stepwise implementation of the virtual factory in manufacturing. Further, [17] propose and show potential of a theoretical framework integrating artificial intelligence (AI), DES and database management technologies. Despite DES simulation being an established tool, the complexity of the modelling means that application of its full potential is still scarce, especially in small and medium-sized enterprises (SMEs) [18]. And there are unresolved issues when applying real-time simulation as short-term decision making [19]. Further, it is established that the process of building DES models is complex i.e., expensive, time consuming, and requires expertise [20, 21]. A considerable combination of quantitative and qualitative skills, large support from many areas of the organization, and extensive knowledge of tools and techniques are required when embarking on DES studies [22].

The outline above raises the possibilities of DES being a method benefiting from the increased emphasis on industrial digitalization and highlight challenges when aiming for DES to become a tool in daily planning of manufacturing and to facilitate the increased digitalized production. This paper reports on a real case study where a Swedish manufacturing company, jointly with a university, investigate the potential of implementation of DES in the company's manufacturing settings. The real case is part of a research project exploring the opportunities and challenges related to applications of simulation, data analysis, and artificial intelligence and human intelligence for future manufacturing.

The start of our study coincided with the first European wave of the Covid-19 pandemic, meaning that the planned activities for action research and case study had to be re-adjusted throughout the study. Thus, we also raise those aspects related to the Covid-19 pandemic in our real case.

The purpose of the study is to investigate the possibilities of applying DES as a tool for decision making in manufacturing to understand the prerequisites for succeeding on such a journey. The ambition is to enhance industrial knowledge of using DES when forming strategies for adoption of new technologies towards increased digitalization.

The research question asked is: *What are the challenges and opportunities for initiation of discrete event simulation as a tool for decision making in I4.0 manufacturing?*

In the following sections related work, concerning the possibilities and aspects of implementation of DES in manufacturing settings, is provided. Subsequently, the present real case is described, followed by findings, discussion, and conclusion.

2 Background and Related Work

This section outlines DES in its historic context and its current and future possible applications related to I4.0 are addressed.

Simulation is defined as an imitation of a system or a real-world process [23] and DES is the modeling of systems in which the state variables change only at discrete set of points in time and is especially useful when simulating systems with variability, interconnectedness, and complexity [21]. Traditionally DES in settings of manufacturing has been applied to study the impact of incorporating e.g., more variants into the production, determine the impact of new equipment or investigating bottleneck scenarios [7, 8]. Another common use is when building a new production facility to create a model on a high level to make a judgment of e.g., production capacity [9]. Further examples of using DES are when determining production planning policies [24, 25]. Meaning that DES is frequently used to support the production system design process for certain scenarios [26] and many applications have been reported in production plant design and in the evaluation of production policies, and planning [27]. The implementation of DES is a process that includes conceptual modelling, data collection and analysis, model coding, experimentation, verification, validation, and confidence [20, 21]. Data collection is often time consuming and there is the implication of availability of data, whether data needs collecting, or if data is not collectable [21]. There are crucial skills necessary when building DES models, such as programming skills, in combination with understanding logistical principles and moreover, awareness of the level of human involvement to determine what data, and how much, is needed for a given purpose [13]. To emphasis, there seems to be a potential of simplifying the process of implementing DES, for example the development of a framework for simulation model simplification, which aims to provide a unifying view, in terms of key activities and enabling and legitimatizing development of educational materials and their uptake [28].

As explained DES has been applied in certain manufacturing decision making, though in general its applications focus on investigating aspects of specific scenarios, often studied separately from daily planning, rather than being used continuously and strategically in long term decision making. Moreover, many companies may not have realized the potential benefits of DES and at the same time the method necessitates specific modelling expertise and requires extended communication between many functions at a company [22]. Further, the DES methodology requires extensive data collection, and lack of expertise not readily available in all businesses may mean that the threshold is too high to engage in implementation of DES. Thus, there may be a lack of understanding of the potentials of DES, especially among SMEs [18]. However, in the era of I4.0 the opportunities of DES have been raised anew as possibilities of real-time data collection, big data analytics coupled with machine learning and its application in short term planning are becoming more realistic [6, 9, 29]. Applications of real-time data driven DES models are becoming a possibility in the era of I4.0 [20, 30]. Implementation of real-time simulation strategies require agile simulation models and short computation times, nonetheless there are indications of the lack of such strategies [19]. The aspects of traditionally time-consuming data collection for DES modelling have the potential to improve in the wake of technologies for real-time data collection [3], though we are not quite there yet [13, 16]. Another aspect is the growing demand for real-time decision making, which evolves DES into a fundamental component of the digital twin, when 'sensing' shop floor data will become vastly available, execution of simulations provide almost real-time solutions enhancing performance of both manufacturing and logistics processes [14]. There is the potential of heightened capability for DES from applying big data analytics at stages of the DES methodology and use of DES in data farming to drive big data analytics techniques [31]. Though connecting DES to real-time data streams and big data sets requires further research [32]. Automatic generation of DES models is a future prospect, though requires further research [9, 33]. Techniques within artificial intelligence, such as applying machine learning [34] or agent-based modelling [35] for decision making can further add to the advancement of DES modelling.

We emphasis that manufacturing companies stress that simulation is an important part of I4.0 [4], indicating the increased interest of such applications in future decision making. However, combining DES with technologies in I4.0 for extended decision making is an area needing further and deeper investigation to reach its potential [9, 13, 16, 32]. It is therefore of interest to explore the challenges and opportunities for DES as a sustainable activity for decision making in increasingly digitalized manufacturing. Meaning there is a need in a manufacturing industry characterized by variability and uncertainty to make the most of I4.0 technologies and find novel ways to support decision making of production planning.

3 Method and Approach

This research has been approached through action research in combination with a real case study. The method of action research *"is driven by a desire to bring about change in practice and it strives toward a form of action in order to identify and solve problems"* [36]. Further, action research is characterized by collaboration between researchers and participants from the setting, where they jointly study and derive solutions to a problem [37]. The research focuses on a real case study, where a researcher and third year bachelor students from a university collaborate with participants from a manufacturing company, to investigate the potential use of DES in production planning decision making. The method of action research, as a bridge between academic research and practical work, focus on practice, change, collaboration, and action [36] and is therefore a suitable approach to investigate the case at hand. Case studies can be exploratory and are suitable when more in-depth knowledge concerning an event is sought-after [36] and when the emphasis is on intensive examination of the setting [37]. The choice of approach and the combination of methods were based on the aspects of close involvement between different actors to jointly investigate the real case to develop and suggest solutions. However, the start of the study coincided with the start of the first European wave of the Covid-19 pandemic. This meant that the planned project and previously experienced models for action research of similar cases had to be re-adjusted throughout the study. Under regular circumstances frequent visits with substantial time spent at the factory site by the university researcher and the bachelor students would have taken place. It is common, in Sweden, for engineering students to mainly be based at the company during their thesis work. As visits to the factory site were strictly regulated because of the Covid-19 pandemic it was only possible for the bachelor students to pay one visit to the company very early on and the researcher could not make any visits to the company. Nor could the researcher, who was supervising the students, meet physically with the students at any time. Those implications meant that new ways of communicating had to be applied. Also, the aspect of understanding a manufacturing facility, with all its processes and production flows, at a distance through mainly studying spread sheets had to be overcome. Nevertheless, throughout the case study, both quantitative and qualitative data were collected from the manufacturing processes at the company. The production data was retrieved from the company Enterprise Resource Planning (ERP) system by the participating project group member from the company. The quantitative data collected was shared via e-mail in the format of Excel files. The qualitative data consisted of meeting notes with company employees to understand the manufacturing layout and processes, and the organization of the company. Regular weekly web-based meetings took place between the three parties, company participant, researcher, and bachelor students. Those meetings were of great importance to understand the manufacturing process and to jointly analyze the data. The qualitative data collection took place in the span between March-June (four months) 2020. The quantitative data collection incorporated data, such as processing times from the manufacturing, which then was analyzed in preparation for building a DES model of the factory shop floor. While the quantitative data was collected during same time period as the qualitative data, the actual data from the quantitative part historically spans over a period of 13,5 months between autumn 2018 until autumn 2019. The collected data was analyzed according to data formats necessary to design and build the DES model.

4 Case Description

The real case is based at a company that manufactures products for the energy sector. There is an interest among middle management to investigate the potential of DES, both to evaluate different production scenarios, but also to explore the potential of this approach becoming a sustainable day-to-day planning activity. No expertise within the area of DES exists at the company, hence the project group is brought together with a researcher from the collaborating university and two third year bachelor students in industrial engineering as part of their final year thesis work. The students have studied courses within logistics and lean manufacturing, though lacks thorough expertise in DES modelling. Hence, they were given access to an online university course on DES when commencing the project. The representative from the company, in the project group, had a positive view of the potential of this method and was engaged throughout the case study and was also the person retrieving the production data from the Enterprise Resource Planning (ERP) system. This proved to be especially vital in this case as the Covid-19 pandemic strongly impacted the communication between project group members. The company has been fairly spared from the Covid-19 pandemic regarding the aspects of not needing to lay off or furlough employees. However, due to the pandemic, the collaboration between the researcher, students, and the company personal was largely affected. It was only possible to visit to the factory site once, very early on, thereafter all other communication took place through web-meetings, e-mails and sharing of Excel-files with production data.

To limit the extent of the data collection for this specific case and to give the company the chance to begin applying DES in a grasping format we jointly identified to focus the study on the company's two highest volume product variants. This limited the scope of the case study, but also gave a base of including substantial data collection for high volume products that passes through many of the different processes on the factory shop floor. The manufacturing processes consist of some CNC-machines and automation equipment, though many processes, for example, within welding are manual. The factory layout is a so-called job shop, meaning that similar equipment or functions are grouped together, such as welding processes located in one area and grinding machines in another. The studied production flows of the two product variants share many resources, though they are routed differently throughout the shop floor. Product variant 1 has 15 operations and Product variant 2 has 19 operations. The production flows are disparate, and some resources are visited several times. The data collection included processing times (planned and reported times), waiting times, repair times (extra operations) and times for sending products on sub-contraction. The data needed thorough categorizing and processing to be analyzed in a suitable format i.e., calculating waiting times and deducting number of repair operations from the data files. The data analysis was time consuming and the group members jointly discussed and categorized the production data, to reach consensus of the interpretation of the data.

5   Results, Analysis and Discussion

This section outlines the findings of the real case study over three aspects; the result of the analyses of the data for building a DES model, results on applying action research and real case study during the Covid-19 pandemic, and the results and learnings made from this real case on how to move forward with DES as regards challenges and opportunities of introducing DES as a tool for decision making in future manufacturing.

5.1   Results from Data Collection Analysis of the Production Data

Based on the data collection, the total production times from starting a new order until the order being completed were calculated. Data was collected to investigate possible discrepancies between planned processing times (retrieved from the ERP system) and reported processing times (reported by operators after completions of operations). Further, from the data we could obtain the number of extra operations (repairs), waiting times between the different operations and variation in sub-contracting times. To clarify the results of the calculations, tables and graphs were drawn for the various aspects studied as outlined above. The Tecnomatix Plant Simulation software [38] was used to model and visualize the production flow according to product routing of the two variants. The production flows are complex with several loops and crossing flows between the two production variants, therefore a simplified DES model was built as part of the case study. Further work and extension of this model is ongoing.

When initiating the study, the company proposed that there might be large discrepancies between the planned and the reported processing times. Though, the study showed that the total discrepancies between planned and reported processing times did not vary as much as anticipated. There was on average only a slight increase in the reported processing times compared to the planned times, and overall the data set followed the same pattern. It should be noted that the planned processing times are set by a production engineer according to a pattern, whereas the reported data is given by the operators themselves after completion of an operation, meaning the reported data may be less reliable, as it is subject to human interaction and judgement.

The company was aware of long waiting times between operations and products are frequently sent to storage in waiting for the next operation. Therefore, there was an interest to study those implications in more detail. The study confirmed a high level of work-in-progress and the extent of the waiting times were quantified, showing the importance of addressing those issues at the case study company.

Early in the production flow the products are sent to a sub-contractor for processing that cannot be performed in-house. Notable the processing times for sub-contracting varied substantially and demonstrated a trend showing how the sub-contracting times increased over the data set studied. Here improved communication with sub-contractors would be beneficial to find causes for this and break the trend of increasing times.

The data analysis showed that for the first product variant studied, 86% of the orders required at least one, but often several repair operations and for the second product variant studied, 43% of orders needed repair operations. The results from the study of repair times indicate that there are quality issues that need further investigation and can be the basis for reformed decision making. Moreover, the added repair operations to the production flow meant that the total processing times increase compared to the originally planned times that initially do not consider or plan for repair cycles. The added repair operations entail an almost constant re-flow of products that revisit operations, making the production flows increasingly complex to analyze and model.

The calculations of the total production times from starting a new order until the order being completed, demonstrated a trend in the data where there is a near 50% increase in total production times, for both product variants, throughout the data sets (see Fig. 1). Due to confidentiality and to avoid disclosure of detailed company data, Fig. 1 does not include the specific production times and unit, though it shall be stressed that the graph displays the result of the analysis of the real production data.

Speculating on the reasons for the increase in production times it can be highlighted that the long waiting times, many repair cycles, and increasingly longer sub-contracting

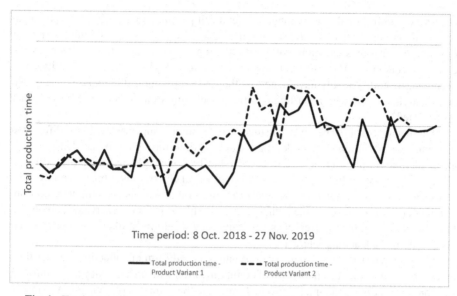

Fig. 1. Total production times for Product Variant 1 and Product Variant 2, respectively.

times have an impact on the total production times. However, the ground reasons why those changes in the data takes place need further investigation as to understand why this is happening and to come to terms with those issues and improve the production flow. The detailed study of the production data showed demonstrated results that need further addressing. In autumn of 2020 the researcher and company representative reached involvement of company management, who showed interest in the results and emphasized importance of continued study. Therefore, the results are now the basis for further investigation leading to future changes in decision making in the real setting.

5.2 Applying Action Research and Case Study During the Covid-19 Pandemic

The study was planned in the months before the first wave of the Covid-19 pandemic hit Europe and when the project started in March 2020 the pandemic was rising fast. Our chosen methods of action research coupled with a real case study therefore quickly needed to be re-adjusted and different ways of communicating, collaborating, and collecting data had to be explored. We had planned frequent visits to spend time at the company to observe and learn the manufacturing processes, collect data, and interview operators on the shop floor. And the most apparent impact on the study was that this was limited to only one visit made by the bachelor students and none by the researcher. Instead, the progression to understand the manufacturing processes and the company problems took place through web-based meetings. During the web-based meetings the company participants explained the manufacturing processes in detail and jointly we extensively studied and analyzed the data files to understand the production flows e.g., times, order of processes, type of equipment, and human resources. This distant way of learning the manufacturing setting was time consuming, repetition was constantly necessary, and it probably took longer to get a full picture and a clear grip of the manufacturing issues. This made the study initially somewhat abstract, and the restrictions to visit the company site meant that the study did not give possibilities for interviewing company personnel. More interactions with operators, logistic planners, quality engineers etc., would probably have benefited the understanding of the manufacturing and i.e., reliability of the data regarding reported processing times could have been further investigated.

Another aspect of not being able to meet was the student supervision. It was difficult to give hands on advice on DES modelling, as this was done through web-based meetings. It also took time for the students to get access to the Tecnomatix Plant Simulation software [38], as licenses were limited to computers in physical classrooms at the university. This was eventually solved by a new system of login through a distance computer solution. The students were given online course materials (films, lectures, exercises booklet) to learn the software, but progress was slow, and they would most likely have progressed faster with more teaching interactions.

The changed ways of communications and adjusted form of collecting data resulted in researcher and students main form of communication with the company being through web-based meetings and e-mail conversations with one company participant. And this in turn made it difficult to disseminate the result of the study at the company as many people were not even aware of the study and meanwhile management were informed there were no clear communication routes or initial interest of the study. Though, in autumn of 2020 the researcher and the company representative found interest from the company management team, meaning the subsequent continuation of the initiative. Management shows a large interest in continuing the work on several levels, both with extended DES models, incorporating more product variants and in parallel to work with improvement of quality issues and enhanced communication with sub-contractors.

The summarized findings from this section highlights the importance of human interaction to interpret, explain and analyze the data, as well as when learning novel software and engaging the interest of management.

5.3   Challenges and Opportunities on a DES Journey Towards I4.0 Manufacturing

As demonstrated the initiation of the DES study and the data collection part have given valuable insights in the details of the production. Moreover, throughout this study, there are lessons learned on a more tacit level related to aspects of introducing DES as a tool for decision making for future manufacturing.

During the data collection we note that manual and human work was required, both for retrieving the data and for the analysis. The data was available in the company's ERP system, though it had to be manually retracted and transferred into Excel files. Data collection and analysis are often described as time consuming, where necessary data is not readily available for collection [21]. Our real case demonstrates and confirms those principles as the data collection and analysis were substantial parts. In this case it can therefore be argued that in real factory settings, data readily available for automatic analysis is yet to become a realistic scenario. Thus, we note the continued interpretation of the data facilitated through human knowledge when collecting, categorizing, analyzing and understanding the data. This interaction with humans in the system also meant that it was difficult to keep a neutral judgement of to the study as preconceptions of anticipated results were raised before and throughout the study. Some of the preconceptions were disproved when analyzing the data, showing how such influence can affect the mindsets of humans. This argues for further possibilities of data collection and analysis with less human interaction in future decision making.

It is known that a range of different competences and specific expertise is necessary when embarking on a DES study [22]. In the real case study presented there was a low knowledge of application of the DES method within the company and the method was newly introduced also to the bachelor students, which emphasized and confirmed the long learning curve to encompass all necessary abilities, i.e., technical, theoretical, quantitative, and qualitative aspects of such a study.

Despite the challenges encountered, the results of the case study demonstrate new knowledge of the production processes that can be brought to light in decision making. The new understanding of the production data is the basis for further studies regarding i.e., improvements of production times, work-in-progress, and highlighting of quality issues. Particularly as the collaborating company is keen to continue with the DES approach to study how production output is affected by the discrepancies in processing and throughput times. Economical key performance indicators related to the aspects of capital bound in work-in-progress is also an aspect the company has highlighted. The study showed the company's immaturity to apply DES and their lack of production data in readily formats for modelling. At the same time the study demonstrated the potential of commencing a DES journey as the results highlight possibilities for improvements of production.

6 Conclusion

The real case outlined in this paper focuses the challenges, yet opportunities for implementation of DES as a tool for decision making in increasingly digital manufacturing. Our results show that the commencement of a DES study, which involves detailed data collection and thorough understanding and analysis of the data, can concretely give input to areas within production that need addressing and require further investigation. This implicates the possibility for revised decision making to improve production times and flows and hence influence production efficiency and improve competitiveness and responding faster to changing markets.

We encountered a variety of challenges during the study. The challenges include the added aspect of performing action research and real case study in midst of the Covid-19 pandemic, the impact from the low level of expertise within the technical aspect of DES modelling, the time-consuming period of data collection and analysis and the necessity of human knowledge and interaction during this process. The emphasis is that it is still vital with human input when understanding and drawing conclusions from data to be the basis for decision making. At the same time, we reflect on the implication that human preconceptions may have on the results, denoting that the data analysis and DES model can be valuable when striving towards objectivity. Consequently, the initiation of the DES approach has increased the understanding of the complexity of the production flows and has endorsed a more holistic view of the factory environment.

Despite the challenges outlined, the results from the real case study have proven a useful insight for the company regarding the outcome of the analysis of production data. The results from the data analysis have been presented and highlighted to the company management team. This in turn has facilitated discussions between managers on how to move forward with fresh studies of production and how to continue with the incentive of implementing DES aiming for improved decision making.

The Covid-19 pandemic meant that adjusted and new ways of communication, collaboration, and data collection were explored and tried. The limitations of communication made it more difficult to disseminate the result of the study at the company and keep management informed. The commencement of a DES journey has demonstrated how the method can act as a catalyst for addressing improvements of production that can lead to more accurate decisions and increase production efficiency. We realize the need for future studies of this area, emphasizing the applicability of DES as tool for decision making and the ambition to enhance industrial knowledge of its possibilities within the era of increased digitalization. We also recognize the importance of human interaction and critical thinking leading Industry 4.0 into Industry 5.0 [39] as our results emphasize human interaction and interpretation throughout the study.

Acknowledgments. The work was carried out at the Production Technology Centre at University West, Sweden and at Siemens Energy AB, Sweden, supported by the Swedish Knowledge Foundation. Their support is gratefully acknowledged.

References

1. Frank, A.G., et al.: Industry 4.0 technologies: implementation patterns in manufacturing companies. Int. J. Prod. Econ. **210**, 15–26 (2019)
2. Machado, C.G., et al.: Industry 4.0 readiness in manufacturing companies: challenges and enablers towards increased digitalization. In: Procedia CIRP 81. 52nd CIRP Conference on Manufacturing Systems, Ljubljana, Slovenia, pp. 1113–1118. Elsevier Ltd. (2019)
3. Alcácer, V., Cruz-Machado, V.: Scanning the industry 4.0: a literature review on technologies for manufacturing systems. Eng. Sci. Technol. Int. J. **22**(3), 899–919 (2019)
4. Narula, S., et al.: Industry 4.0 adoption key factors: an empirical study on manufacturing industry. J. Adv. Manag. Res. **17**(5), 697–725 (2020)
5. Lee, J., et al.: A cyber-physical systems architecture for industry 4.0-based manufacturing systems. Manuf. Lett. **3**, 18–23 (2015)
6. Gasjek, B., et al.: Using maturity model and discrete-event simulation for industry 4.0 implementation. Int. J. Simul. Modell. **18**(3), 488–499 (2019)
7. Andrade-Gutierrez, E.S., Carranza-Bernal, S.Y., Hernandez-Sandoval, J., Gonzalez-Villarreal, A.J., Berber-Solano, T.P.: Optimization in a flexible die-casting engine-head plant via discrete event simulation. Int. J. Adv. Manuf. Technol. **95**(9–12), 4459–4468 (2018). https://doi.org/10.1007/s00170-017-1562-9
8. Faget, P., et al.: Applying discrete event simulation and an automated bottleneck analysis as an aid to detect running production constraints. In: Kuhl, M.E., Steiger, N.M., Armstrong, F.B., Joines, J.A. (eds.) Proceedings of the 2005 Winter Simulation Conference, WSC 2005, Orlando, FL, USA (2005)
9. Viera, A.A.C., et al.: Setting an industry 4.0 research and development agenda för simulation – a literature review. Int. J. Simul. Modell. **17**(3), 377–390 (2018)

10. Timm, I.J., Lorig, F.: Logistics 4.0 – a challenge for simulation. In: Yilmaz, L., Chan, W.K.V., Moon, I., Roeder, T.M.K., Macal, C., Rossetti, M.D. (eds.) Proceedings of the 2015 Winter Simulation Conference, WSC 2015, Huntington Beach, CA, USA (2015)

11. Lidberg, S., Aslam, T., Pehrsson, L., Ng, A.H.C.: Optimizing real-world factory flows using aggregated discrete event simulation modelling. Flex. Serv. Manuf. J. **32**(4), 888–912 (2019). https://doi.org/10.1007/s10696-019-09362-7

12. Jung, W.-K., et al.: Real-time data-driven discrete-event simulation for garment production lines. Prod. Plann. Control, 1–12 (2020). https://doi.org/10.1080/09537287.2020.1830194

13. Gershwin, S.B.: The future of manufacturing systems engineering. Int. J. Prod. Res. **56**(1–2), 224–237 (2018)

14. Agalianos, K., et al.: Discrete event simulation and digital twins: review and challenges for logistics. In: Procedia Manufacturing. 30th International Conference on Flexible Automation and Intelligent Manufacturing FAIM 2021, Athens, Greece, vol. 51, pp. 1636–1641. Elsevier Ltd. (2020)

15. Raj, A., et al.: Barriers to the adoption of industry 4.0 technologies in the manufacturing sector: an inter-country comparative perspective. Int. J. Prod. Econ. **224**, 107546 (2020)

16. Dalstam, A., et al.: A stepwise implementation of the virtual factory in manufacturing industry. In: Rabe, M., Juan, A.A., Mustafee, N., Skoogh, A., Jain, S., Johansson, B. (eds.) Proceedings of the 2018 Winter Simulation Conference, WSC 2018, Gothenburg, Sweden, pp. 3229–3240 (2018)

17. Moges Kasie, F., et al.: Decision support systems in manufacturing: a survey and future trends. J. Model. Manag. **12**(3), 432–454 (2017)

18. Sobottka, T., et al.: A case study for simulation and optimization based planning of production and logistics systems. In: Chan, W.K.V., D'Ambrogio, A., Zacharewicz, G., Mustafee, N., Wainer, G., Page, E. (eds.) Proceedings of the 2017 Winter Simulation Conference, WSC 2017, Las Vegas, NV, USA (2017)

19. Lugaresi, G., Matta, A.: Real-time simulation in manufacturing systems: challenges and research directions. In: Rabe, M., Juan, A.A., Mustafee, N., Skoogh, A., Jain, S., Johansson, B. (eds.) Proceedings of the 2018 Winter Simulation Conference, WSC 2018, Gothenburg, Sweden (2018)

20. Law, A.M., Kelton, W.D.: Simulation Modeling and Analysis, 3rd edn. McGraw-Hill, New York (2000)

21. Robinson, S.: Simulation the Practice of Model Development and Use, 2nd edn. Palgrave Macmillan, New York (2014)

22. Sturrock, D.T.: Tips for successful practice of simulation. In: Jain, S., Creasey, R.R., Himmelspach, J., White, K.P., Fu, M. (eds.) Proceedings of the 2011 Winter Simulation Conference, WSC 2011, Phoenix, AZ, USA (2011)

23. White, K.P., Ingalls, R.G.: Introduction to simulation. In: Yilmaz, L., Chan, W.K.V., Moon, I., Roeder, T.M.K., Macal, C., Rossetti, M.D. (eds.) Proceedings of the 2015 Winter Simulation Conference, WSC 2015, Huntington Beach, CA, USA (2015)

24. Zandieh, M., Motallebi, S.: Determination of production planning policies for different products in process industries: using discrete event simulation. Prod. Eng. Res. Devel. **12**(6), 737–746 (2018). https://doi.org/10.1007/s11740-018-0843-y

25. Kibira, D., et al.: Framework for standardization of simulation integrated production planning. In: Roeder, T.M.K., Frazier, P.I. , Szechtman, R., Zhou, E., Huschka, T., Chick, S.E. (eds.) Proceedings of the 2016 Winter Simulation Conference, WSC 2016, Washington, DC, USA, pp. 2970–2981 (2016)

26. Flores-Garcia, E., et al.: Simulation in the production system design process of assembly systems. In: Yilmaz, L., Chan, W.K.V., Moon, I., Roeder, T.M.K., Macal, C., Rossetti, M.D. (eds.) Proceedings of the 2015 Winter Simulation Conference, WSC 2015, Huntington Beach, CA, USA, pp. 2124–2135 (2015)

27. Semini, M., et al.: Applications of discrete-event simulation to support manufacturing logistics decision-making: a survey. In: Perrone, L.F., Wieland, F.P., Liu, J., Lawson, B.G., Nicol, D.M., Fujimoto R.M. (eds.) Proceedings of the 2006 Winter Simulation Conference, WSC 2006, Monterey, CA, pp. 1946–1953 (2006)

28. van der Zee, D.-J.: Model simplification in manufacturing simulation – review and framework. Comput. Ind. Eng. **127**, 1056–1067 (2019)

29. Winkelhaus, S., Grosse, E.H.: Logistics 4.0: a systematic review towards a new logistics system. Int. J. Prod. Res. **58**(1), 18–43 (2020)

30. Gao, Y., et al.: Real-time modelling and simulation method of digital twin production line. In: IEEE 8th Joint International Information Technology and Artificial Intelligence Conference (ITAIC) (2019)

31. Greasley, A., Edwards, J.S.: Enhancing discrete-event simulation with big data analytics: a review. J. Oper. Res. Soc. **25**(6), 534–548 (2019)

32. Turner, C.J., et al.: Discrete event simulation and virtual reality use in industry: new opportunities and future trends. IEEE Trans. Hum.-Mach. Syst. **46**(5), 2168–2291 (2016)

33. Reinhart, H., et al.: A survey on automatic model generation for material flow simulation in discrete event manufacturing. Procedia CIRP **81**, 121–126 (2019)

34. Nagahara, S., et al.: Toward data-driven production simulation modeling: dispatching rule identification by machine learning techniques. Proceedia CIRP **81**, 222–227 (2019)

35. Fauzan, A.C., et al.: Simulation of agent-based and discrete event for analysing multi organisational performance. In: International Seminar on Application for Technology of Information and Communication (iSematic), Semarang, Indonesia, 21–22 September 2019 (2019)

36. Säfsten, K., Gustavsson, M.: Research Methodology – For Engineers and Other Problem-Solvers. Studentlitteratur, Lund (2019)

37. Bryman, A.: Social Research Methods, 5th edn. Oxford University Press, Oxford (2016)

38. Bangsow, S.: Tecnomatix Plant Simulation – Modeling and Programming by Means of Examples, 2nd edn. Springer, Cham (2020). https://doi.org/10.1007/978-3-030-41544-0

39. Javid, M., Haleem, A.: Critical components of Industry 5.0 towards a successful adoption in field of manufacturing. J. Integr. Manag. **3**, 327–348 (2020)

Methodology for Multi-aspect Ontology Development

Use Case of DSS Based on Human-Machine Collective Intelligence

Alexander Smirnov(ID), Tatiana Levashova(✉)(ID), Andrew Ponomarev(ID),
and Nikolay Shilov(ID)

St. Petersburg Federal Research Center of the Russian Academy of Sciences, 39, 14th Lineline,
St. Petersburg, Russia
{smir,tatiana.levashova,ponomarev,nick}@iias.spb.su

Abstract. The lack of interoperability observed in modern DSSs becomes even greater when complex systems covering multiple domains are considered. In the present research, the apparatus of multi-aspect ontologies is used as a means to represent knowledge of DSSs based on human-machine collective intelligence for enabling interoperability between the system components and coordinate interrelated processes. The available ontology development methodologies are not quite suitable for the development of multi-aspect ontologies because they leave aside the problem of choosing approaches for integration of reusable ontologies. Since the structure of a multi-aspect ontology imposes some restrictions on the aspects integration, the objective of this research is to propose a methodology for the development of multi-aspect ontologies that incorporates an aspects integration approach. For the research purpose, the existing ontology development methodologies have been analyzed and an ontology development pattern followed by most methodologies has been revealed. The developed four-stage methodology extends it with an aspects integration approach. The methodology is illustrated through the development of a multi-aspect ontology to support semantic interoperability in DSSs based on human-machine collective intelligence.

Keywords: Ontology development methodology · Multi-aspect ontology ·
Decision support · Human-machine collective intelligence

1 Introduction

Collective intelligence is an emergent property from the synergies among data/information/knowledge, software/hardware, and humans with insight that continually learns from feedback to produce just-in-time knowledge for better decisions than any of these elements acting alone [1]. A collective intelligence system, which connects these three elements into a single interoperable platform, is believed to improve the efficiency of decision support.

At the moment, decision support systems (DSSs) that are based on human-machine collective intelligence (HMCI) are not numerous. Few DSSs of this type do not leverage the full potential of HMCI. Basically, they support decision-making by human groups

© Springer Nature Switzerland AG 2021
U. Jayawickrama et al. (Eds.): ICDSST 2021, LNBIP 414, pp. 97–109, 2021.
https://doi.org/10.1007/978-3-030-73976-8_8

providing them with a specially developed software (e.g., [2, 3]) and teamwork between humans and machines has not been achieved so far [4].

In DSSs based on HMCI, various processes that relate to different domains (e.g., knowledge management, decision-making, etc.) are ongoing and knowledge from multiple domains is required to support decision-making in such systems. In these systems, humans and machines must interoperate so that they could exchange their views on problems, discuss alternatives, make agreements, etc. The lack of interoperability observed in modern DSSs [5–7] becomes even greater when complex systems covering multiple domains are considered. The apparatus of multi-aspect ontologies was proposed as a means to represent knowledge of complex systems, enable interoperability between their components, and coordinate interrelated processes [8].

A multi-aspect ontology comprises three levels: local, aspect, and global. The local level represents concepts and relationships observed only from one view. Each aspect can be represented by a specific formalism. The aspect level represents concepts and relationships from the local level that are shared by two or more aspects. The aspect level defines the formalism of the multi-aspect ontology. The global level is the common part of the multi-aspect ontology represented using the multi-aspect ontology formalism. The concepts represented at this level are related to those of the aspect level.

Although existing ontology development methodologies imply ontology reuse and integration to build a new ontology, the choice of ontology integration approach is left to the developers. The multi-aspect ontologies rely upon a certain structure to support integration of heterogeneous local aspects. This structure imposes some restrictions on the integration approach, which means that developing multi-aspect ontologies requires a methodology that would incorporate an approach to aspects integration. The contribution of this paper is a methodology for the development of multi-aspect ontologies and a resulting multi-aspect ontology aimed to support semantic interoperability between humans and machines in an HMCI-based DSS, which has been developed following the proposed methodology. Unlike related approaches to ontology integration (e.g., [9–11]), the approach proposed in the methodology does not require a prior agreement on a global ontology level. Instead, this level is a result of aspects integration.

The rest of the paper is organized as follows. Section 2 contains some basic information about the HMCI environment, explaining the scope of the multi-aspect ontology application. The methodology for the development of multi-aspect ontologies and its application are discussed in Sect. 3. The main research results are discussed in the Conclusion.

2   Decision Support Based on Human-Machine Collective Intelligence

In the present research, a collective intelligence environment supports joint work of relatively small and short-living (hours to several days) ad hoc teams of humans and machines on decision support problems allowing participants to self-organize (define and adapt the plan of actions). In particular, the HMCI environment provides a basic mechanism of communication and coordination between the heterogeneous participants (humans and machines), allowing information exchange on two levels: information concerning the problem being solved (available data, opinions, arguments and models), and process information (role distribution, responsibilities and so on) [12].

The following principal actors are distinguished in the environment: end-user (decision-maker), participant, and service provider. End-user (decision-maker) uses the environment to get help in making a decision. He/she describes the problem and posts so that the problem description is visible to a specified community. Participants (humans and/or software services) self-organize into teams to work on the problem given by the end-user. Finally, the service provider develops, integrates to the environment, and supports software services that can act on behalf of participants working on some problem given by the end-user.

Two processes take place when a team consisting of humans and software services are working on a problem: search for a solution and decision support (re)organization. Searching for a solution is a main productive process, during which team members enrich the problem initially defined by the end-user with new information. The general flow of this process is driven by decision-making methodologies and the decision-making model introduced by H. Simon. Decision support (re)organization process represents all the activities aimed at planning and organization of teamwork (e.g., deciding whether additional resources are required, assigning team member responsibilities, giving deadlines for completing a task, and identifying new tasks to be solved for reaching the goals of the whole process). To support the (re)organization process the environment suggests process assistance mechanisms that offer efficient modes and scenarios of the collaboration depending on the status of the problem-solving process.

Technologically, the processes performed by the team are organized with a help of ontology-based smart space, accessible by software services (via SPARQL-based APIs) and human participants (via GUI components, hiding technical details of ontology usage). This smart space contains the structured description of the problem, gradually extended and refined by the team. So, the progress of a team on the problem is reflected by structural changes of an ontological description of the problem.

3   Methodology for Development of Multi-aspect Ontology

An analysis of ontology development methodologies underlies the methodology for the development of multi-aspect ontology to support semantic interoperability in decision support systems based on human-machine collective intelligence. "Enterprise" [13], TOVE [14], METHONTOLOGY [15], Protégé [16, 17], On-to-Knowledge [18], NeOn [19], Lifecycles [20], and AMOD [21] are among the analyzed methodologies. These methodologies follow approximately the same ontology development pattern: requirements specification, creation of conceptualization, conceptualization formalization, ontology implementation, and ontology evaluation. A methodology of the multi-aspect ontology development adopts this pattern.

3.1   Multi-aspect Ontology Development

The multi-aspect ontology development goes through four stages (Fig. 1).

The first stage aims at producing of ontology requirements specification, the identification of the purpose and scope of the ontology, and the identification of kinds of
aspects to be included into the multi-aspect ontology. The methodology of the multi-
aspect ontology development does not impose special demands on methods of requirements specification. Nevertheless, most of the ontology development methodologies
recommend competency questions for this objective. The requirements specification is
the basis to identify the purpose and scope of the ontology. These purpose and scope
provide ideas about the kinds of aspects.

The second stage focuses on the development of aspect ontologies. The stage starts
with the requirements specification for each aspect. Then, two scenarios of the aspect
ontologies development can be used: development from scratch or ontology reuse.

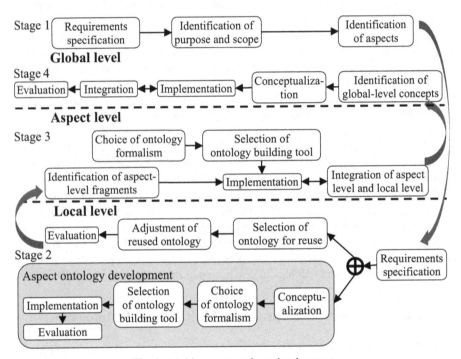

Fig. 1. Multi-aspect ontology development

Ontology reuse is considered to be of a higher priority. The reused ontologies are
adjusted to the aspect ontologies requirements without changing the original formalism
of those ontologies. The adjusted ontologies are evaluated. In respect that the developers
of the reused ontologies evaluated them, the evaluation concerns checking the consistency of the representations for the concepts and properties that have been changed while
the adjustment. These representations must correspond to the representations used in the
source ontologies. Here, consistency of representations means correct spelling of names
for the concepts and properties, consistency of their grammatical forms, lack of redun-
dancy, etc. In details, the Protégé methodology [17] describes this kind of evaluation; it
is referred to as verification there.

If no ontologies for reuse have been found, the aspect ontology is developed from scratch in accordance with the ontology development pattern adhered by most ontology development methodologies: requirements specification, conceptualization, formalization, implementation, and evaluation. Due to a number of tools supporting ontology building have been created to date, the methodology for multi-aspect ontology development addresses the issues of formalization and implementation jointly. The ontology building tools support the aspect ontology formalization in a chosen formalism or representation language and the ontology encoding, i.e. its implementation. The result of the implementation is a logical model derived from the aspect conceptualization using an ontology building environment. The resulting model is verified for the representation consistency and evaluated for the logical consistency. The logical consistency is checked by an ontology reasoner integrated into the ontology building environment.

The third stage targets representing the aspect level of the ontology. At this stage, fragments of the aspect ontologies are identified that represent concepts of the aspect level and relationships between them. As the aspect level concepts, we capture concepts for which there are semantic mappings to the concepts of one or more aspect ontologies.

The ontology fragments are represented in the same formalism (hereinafter referred to as the multi-aspect ontology formalism). Generally, this formalism does not depend on the formalisms of the (local) aspect ontologies. Multi-aspect ontology formalism and an ontology building tool to represent the fragments can be chosen before, in parallel with, or after the development of the aspect ontologies. It depends on the developers' goals. If the local level represents many aspect ontologies implemented using the same formalism, sometimes it makes sense to formalize the aspect level in the formalism of those ontologies. However, this recommendation is not mandatory.

Evaluation of the aspect-level fragments consists in checking whether the conceptualization formalized by a fragment complies with the conceptualization of the aspect ontology out of that this fragment has been captured.

The final activities at the third stage concern integration of the aspect and local levels. They are integrated via unambiguous aligning the aspect-level fragments and the aspect ontologies. The alignment relationships are formalized by means of aspect ontology formalisms.

The fourth stage is intended for the development of the global level of the ontology. For this, concepts common for several aspect-level fragments are identified. Concepts of the global level are selected out of the common concepts and then a conceptualization of the global level is created. This conceptualization is formalized and implemented using an ontology building tool (usually, the tool is chosen at stage 3).

The global level and the aspect level are integrated via bridging rules [8] that are specified between the global level concepts and the concepts of the aspect-level fragments. The bridging rules are formalized by means of the multi-aspect ontology formalism. Bidirectional arrows in Fig. 1 between the implementation and integration blocks indicate that the activities on the ontology implementation and ontology integration are interrelated. In particular, one can either integrate knowledge when creating a conceptualization, and then implement this conceptualization, or introduce (integrate) some knowledge into an implemented ontology and therefore to change the original conceptualization.

Activity on the evaluation of the multi-aspect ontology finalizes the fourth stage. The evaluation implies checking the global level representation consistency, checking for the lack of redundant relationships at this level, and checking the logical consistency of the aspect and global levels. At the fourth stage, the logical consistency is checked via passing facts between the aspect and global levels using the bridging rules.

Relationships between the levels of the multi-aspect ontology are explained by an example of the appearance of an individual at the local level. For simplicity, it is supposed that the individual belongs to a class that is represented at the aspect level. When such an individual appears at the local level, this individual is unambiguously translated (by the mechanisms of the local aspect ontology) to the aspect level due to the alignment between the classes of these levels. Then, the ontology reasoner that supports the global level classifies the individuals based on bridging rules. Finally, again by means of the bridging rules, classes of the aspect level are instantiated and, in turn, are unambiguously translated to the local level.

3.2   Use Case of DSS Based on Human-Machine Collective Intelligence

The development of a multi-aspect ontology to support semantic interoperability in DSSs based on human-machine collective intelligence follows the stages proposed by the methodology introduced above.

Stage 1. Global level.

Specification of requirements to the multi-aspect ontology of the HMCI environment. The purpose and scope of the multi-aspect ontology, and competency questions form the basis to produce ontology specification. This specification comprises a set of initial concepts considered relevant to the modelled domain. With reference to the global level, these concepts correspond to kinds of aspects viewed in the ontology.

The purpose of the multi-aspect ontology is the support of semantic interoperability of components of a decision support system based on human-machine collective intelligence. Relevant concepts revealed from the ontology purpose definition are *component, human-machine collective intelligence, decision support.*

The ontology scope is the self-organization of software services and humans based on their competencies to solve the user task as a decision support problem. Relevant concepts revealed from the ontology scope definition are *software service, human, self-organization, competency, decision support, task,* and *problem.*

Competency questions:

- How is ensured semantic interoperability of heterogeneous participants of the human-machine environment?
- What knowledge should the ontology represent to enable decision support?
- What knowledge should the ontology represent to enable self-organization?

The semantic interoperability of heterogeneous participants of the human-machine environment is ensured by concepts of the global ontology level. Relevant concepts here are *participant* and *human-machine environment.* The participants self-organize to enable decision support. Relevant concepts are *participant, self-organization, decision support*. Software services and humans self-organize based on their competencies. Relevant concepts are *software service, human, self-organization, competency*.

Identification of aspects to be included into the multi-aspect ontology. Ontology aspects are chosen based on the revision of the set of initial concepts. The revision concern the identification of possible synonyms, subconcepts, irrelevant concepts or lacking concepts. The revision showed that regarding the human-machine environment *component* and *participant* are synonyms. These concepts are decided to be referred to as *participant*. At the same time, *participant* is a subconcept for the concept of *human-machine environment* and therefore there is no need to represent *participant* as a separate aspect. The concepts of *software service* and *human* are kinds of *participants* and consequently are not represented as aspects. The *user task* in terms of decision support is considered as *problem*, that is *problem* and *task* are synonyms. Term *task* is chosen to name these concepts. The concept *task* belongs to the *decision support* domain and is used in conjunction with knowledge of the *subject domains* that supply the input data for this task. *Collective intelligence* is the product of activities of the human-machine environment participants and cannot be considered as an aspect.

Summing up, the following aspects are identified that have to be introduced into the multi-aspect ontology: *decision support, subject domain, human-machine environment, self-organization*, and *competency*. Due to space restriction, the further discussion focuses on aspects of *decision support* and *human-machine environment*.

Stage 2. Local level.

The ontology specification for the *decision support* aspect is made based on the definitions for the ontology purpose and scope, and the analysis of decision-making methodologies [22]. The specification for the aspect of *human-machine environment* is defined in the same way as the specification of the multi-level ontology (ontology purpose, scope, competency questions).

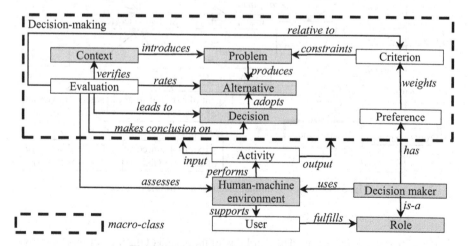

Fig. 2. Ontology for decision support aspect

Decision Support. Decision support ontologies [23, 24] are reused to develop the ontology of the decision support aspect. Figure 2 presents this ontology. It comprises the following concepts: • *Human-machine environment* – an environment comprising humans and machines which jointly solve the problem formulated by the user as a decision support problem; • *Context* – information that can be used to characterize the decision situation; context represents reasons, facts, contradictions, and other situation-related information; • *Problem* – an issue that has to be resolved by finding an answer to it or taking some actions; • *User* – one who uses the human-machine environment as a decision support tool; • *Decision maker* – one who makes the final choice among the alternatives; • *Role* – decision support and decision-making activities required or expected of the users and participants of the human-machine environment within the role; • *Criterion* – a rule or standard by which alternatives can be ranked based on the decision-maker preferences; • *Preference* – objective function and the desired value of the objective function; • *Alternative* – optional problem solutions or courses of action; • *Decision* – agreement to adopt an alternative to resolve the problem; • *Evaluation* – judgment how much the object being evaluated (the context, the human-machine environment, the decision) addressed what is expected; • *Activity* – actions related to achievement any goal while decision support (e.g., information gathering, context analysis, development of alternatives, etc.); • *Decision-making* – a process that is used to make a decision being solution to the problem.

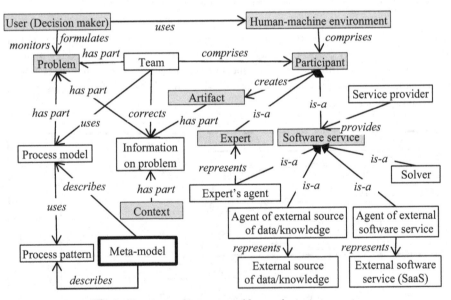

Fig. 3. Ontology of human-machine environment aspect

Human-Machine Environment. The ontology for the aspect of the human-machine environment is developed from scratch. The ontology is given in Fig. 3. It comprises the following concepts: • *Human-machine environment* – a community of software services and machines the interactions of which leads to the emergence of collective intelligence; • *User (Decision maker)* – one who formulates the problem to the human-machine environment, monitors the progress of its solution, makes adjustments to the solution process and a final decision based on the information provided by the environment; • *Problem* – an issue that has to be resolved by finding an answer to it or taking some actions; • *Participant* – software service or human taking part in decision support; • *Software service* – software provided as service; • *Service provider* – a person or a company proving the services to the subscribers; • *Expert* – human as a participant of the human-machine environment; • *Expert's agent* – software service acting on behalf of the expert; • *Team* – a set of participants implementing the model of decision support process for a specific problem; • *Artifact* – a result created by a participant as an outcome of the participant's activity; • *Agent of an external source of data/knowledge* – software service as a participant that provides access to an external source of data/knowledge; • *External source of data/knowledge* – a ready-to-use source of data or knowledge (e.g., a database, a service) that is hosted, supported and maintained by a source provider and to which the environment has access under specified conditions; • *Agent of external software service* – software service as a participant that provides access to an external software service; • *External software service* – ready-to-use software service that is hosted, supported and maintained by a service provider and to which the environment has access under specified conditions; • *Solver* – intelligent software that synthesizes information and knowledge to achieve a solution for subproblems appeared while the participant activities on decision support; • *Process model* – a description of the processes on data collection, investigation, and alternative evaluation created using the meta-model; • *Meta-model* – a set of elements for the creation of a given process model; • *Process pattern* – a typical decision process for solving routine tasks within coordination and self-organization; • *Information on problem* – context, information on the problem from the user, and information on the problem produced by the team in the course of its activities on searching a solution for the problem; • *Context* – any information that characterizes the situation of an entity where an entity can be place, a participant of the human-machine environment, or the user.

Stage 3. Aspect level.
The shadowed blocks in Fig. 2 and Fig. 3 represent concepts that are chosen as the aspect-level concepts. Table 1 summarizes these concepts. The concepts of the aspect level and the local level are aligned correspondingly.

The OWL ontology description language [25] is used for the aspect level formalization since it is the most widely used up-to-date means supported by the World Wide Web Consortium (W3C). Therefore, OWL becomes the representation language for the multi-aspect ontology. The Protégé ontology editor and building framework [26] is used to implement the aspect level.

Stage 4. Global level.
The list of concepts that are common for several aspects is as follows: *human-machine environment, software service, human, participant, decision maker, role, decision, artifact, task,* and *context.* Referring to the aspects of *self-organization* and *competency,*

Table 1. Concepts of the aspect level.

Local-level aspect	Aspect-level concept
Decision support	Human-machine environment, user, decision maker, role, problem, context, alternative, decision
Human-machine environment	Human-machine environment, user (decision maker), participant, software service, expert, problem, artifact, context

which are not considered in the paper, the representations of those aspects include concepts from the list and the list of the common concepts does not contain concepts specific for those aspects.

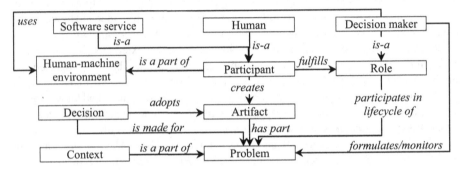

Fig. 4. Global level of multi-aspect ontology

The global level of the multi-aspect otology is presented in Fig. 4. The correspondences between the concepts of the aspect level and global level for which bridging rules have to be formalized are shown in Table 2.

The global level is formalized in OWL. The logical consistency of the global level is evaluated using the Pellet reasoner provided by the Protégé framework.

The following illustrative example based on the excerpt of the implemented multi-aspect ontology demonstrates the applicability of this ontology to decision support situations. During the problem-solving process, the participants of the human-machine environment (Fig. 3) generate various *artifacts* (the concept *"Artifact"*). In fact, the artifact generation process is a part of the decision support process (Fig. 2), and the generated *artifacts* are at the same time *alternatives* (the concept *"Alternative"*) at this stage of decision support. A bridging rule formalizes equivalence of the global-level concept *"Artifact"* and the concept *"Alternative"* (Table 2). Thus, generating artifacts, the human-machine environment at the same time generates decision alternatives.

With respect to application domains, the multi-aspect ontology represents knowledge of an application domain as an aspect, and depending on the particular problem from a given domain, this aspect can be replaced with another one on the "plug-in" basis. Let us consider the domains of "e-tourism" and "smart city". For instance, in the former domain, the problem of finding the most interesting attractions in the area meeting the

Table 2. Correspondences between concepts of aspect level and global level.

Global level	Decision support	Human-machine environment
Human-machine environment	Human-machine environment	Human-machine environment
Software service	Software service	Software service
Human	–	Expert
Participant	User	Participant
Decision maker	Decision maker	User (decision maker)
Role	Role	Expert
Decision	Decision	Artifact
Artifact	Alternative, decision	Artifact
Problem	Problem	Problem
Context	Context	Context

tourist's preferences is being solved. Here, at the stage of alternatives development, the *artifact* corresponds to the concept "*attraction*" and is a name and description of an attraction. In the latter domain, the problem of finding a way to get from city location A to city location B is being considered. In this case, the *artifact* corresponds to the concept "*route*", describing the transportation means, stops, and transfers. As a result, the multi-aspect ontology becomes problem-specific, however, all other aspects remain unchanged.

4 Conclusion

A four-stage methodology for the development of multi-aspect ontologies is proposed. Multi-aspect ontologies are used as a means to represent knowledge of DSSs based on HMCI, enable interoperability in them, and coordinate interrelated processes ongoing in such DSSs. The methodology adopts an ontology development pattern followed by most ontology development methodologies and extends these methodologies with an ontology integration approach. Although this approach supposes the usage of a shared global ontology, this ontology is not imposed as conformance to a common standard, but it is the result of the integration of ontologies representing multiple domain knowledge from which is combined with the purpose of decision support. The multi-aspect ontology allows an HMCI-based DSS to easily adapt to new application domains due to the global level that represents common concepts for any DSSs of this type and the possibility to incorporate application knowledge as a local aspect, followed by the mapping this knowledge and the knowledge represented at the aspect level not changing any other relationships between the ontology levels. The methodology is beneficial for complex systems dealing with knowledge from multiple domains; it enables developing shared ontologies and does not require an agreement on a top-level ontology.

The main drawback of the methodology concerns the manual ontology mapping. At the same time, for the application knowledge it is the only way so far, since kinds of mappings between the application aspect and the multi-aspect ontology levels depend on the decision support problem and each problem requires specific mappings. Nevertheless, knowledge represented at other levels can be mapped. But the manual mapping requires significant efforts and existing automatic mapping techniques produce high-quality results only in narrow domains. The proposed approach seems to be an efficient solution since it reduces the need for mapping but at the same time enables involvement of automated mapping to aspects that represent narrow domains.

Acknowledgement. The research is funded by the Russian Science Foundation (project no. 19-11-00126).

References

1. Glenn, J.C.: Collective intelligence and an application by the millennium project. World Futur. Rev. **5**, 235–243 (2013). https://doi.org/10.1177/1946756713497331
2. Zhang, Y.: CityMatrix – an urban decision support system augmented by artificial intelligence, MS thesis, Massachusetts Institute of Technology (2017). https://dam-prod.media.mit.edu/x/2017/08/23/ryanz-ms-17.pdf. Accessed 08 Feb 2021
3. Willcox, G., Rosenberg, L., Askay, D., Metcalf, L., Harris, E., Domnauer, C.: Artificial swarming shown to amplify accuracy of group decisions in subjective judgment tasks. In: Arai, K., Bhatia, R. (eds.) FICC 2019. LNNS, vol. 70, pp. 373–383. Springer, Cham (2020). https://doi.org/10.1007/978-3-030-12385-7_29
4. Berditchevskaia, A., Baeck, P.: The future of minds and machines. Nesta Report. London (2020). https://media.nesta.org.uk/documents/FINAL_The_future_of_minds_and_machines.pdf. Accessed 08 Feb 2021
5. Pease, S.G., et al.: An interoperable semantic service toolset with domain ontology for automated decision support in the end-of-life domain. Futur. Gener. Comput. Syst. **112**, 848–858 (2020). https://doi.org/10.1016/j.future.2020.06.008
6. Roosan, D., Hwang, A., Law, A.V., Chok, J., Roosan, M.R.: The inclusion of health data standards in the implementation of pharmacogenomics systems: a scoping review. Pharmacogenomics **21**, 1191–1202 (2020). https://doi.org/10.2217/pgs-2020-0066
7. Zacharewicz, G., Daclin, N., Doumeingts, G., Haidar, H.: Model driven interoperability for system engineering. Modelling. **1**, 94–121 (2020). https://doi.org/10.3390/modelling1020007
8. Smirnov, A., Levashova, T., Shilov, N., Ponomarev, A.: Multi-aspect ontology for interoperability in human-machine collective intelligence systems for decision support. In: Proceedings of the 11th International Joint Conference on Knowledge Discovery, Knowledge Engineering and Knowledge Management, pp. 458–465. SCITEPRESS - Science and Technology Publications (2019). https://doi.org/10.5220/0008356304580465
9. de Silva, P.A., Ribeiro, C.M.F.A., Schiel, U.: Formalizing ontology reconciliation techniques as a basis for meaningful mediation in service-related tasks. In: Proceedings of the ACM First Ph.D. Workshop in CIKM, pp. 147–154. ACM Press, New York (2007). https://doi.org/10.1145/1316874.1316898
10. Boulkroun, B., Benchikha, F., Bachtarzi, C.: Integrating ontological data sources using viewpoints-based approach. J. Comput. Inf. Technol. **24**, 383–400 (2016). https://doi.org/10.20532/cit.2016.1003228
11. Pai, F.-P., Yang, L.-J., Chung, Y.-C.: Multi-layer ontology based information fusion for situation awareness. Appl. Intell. **46**(2), 285–307 (2016). https://doi.org/10.1007/s10489-016-0834-7

12. Smirnov, A., Ponomarev, A.: Decision support based on human-machine collective intelligence: major challenges. In: Galinina, O., Andreev, S., Balandin, S., Koucheryavy, Y. (eds.) NEW2AN/ruSMART-2019. LNCS, vol. 11660, pp. 113–124. Springer, Cham (2019). https://doi.org/10.1007/978-3-030-30859-9_10

13. Uschold, M., King, M.: Towards a methodology for building ontologies. In: Proceedings of the Workshop on Basic Ontological Issues in Knowledge Sharing. International Joint Conference on Artificial Intelligence (1995). 10.1.1.697.8733

14. Grüninger, M., Fox, M.S.: Methodology for the design and evaluation of ontologies. In: Proceedings of the IJCAI-95 Workshop on Basic Ontological Issues in Knowledge Sharing (1995)

15. Fernandez, M., Gomes-Perez, A., Juristo, N.: METHONTOLOGY: from ontological art towards ontological engineering. In: AAAI Proceedings of the Symposium on Ontological Engineering, pp. 33–40. AAAI (1997)

16. Noy, N.F., McGuinness, D.L.: Ontology development 101: a guide to creating your first ontology. Stanford University Report, Stanford (2001). https://www.ksl.stanford.edu/people/dlm/papers/ontology-tutorial-noy-mcguinness.pdf. Accessed 08 Feb 2021

17. Noy, N.F., Tu, S.: Ontology Engineering for the Semantic Web and Beyond (2003). https://slideplayer.com/slide/12403284/. Accessed 08 Feb 2021

18. Sure, Y., Staab, S., Studer, R.: On-to-knowledge methodology (OTKM). In: Staab, S., Studer, R. (eds.) Handbook on Ontologies, pp. 117–132. Springer, Heidelberg (2004). https://doi.org/10.1007/978-3-540-24750-0_6

19. Suárez-Figueroa, M.C., Gómez-Pérez, A., Fernández-López, M.: The NeOn methodology framework: a scenario-based methodology for ontology development. Appl. Ontol. **10**, 107–145 (2015). https://doi.org/10.3233/AO-150145

20. Neuhaus, F., Vizedom, A.: Ontology Summit 2013 Communiqué: Towards Ontology Evaluation across the Life Cycle (2013). https://ontolog.cim3.net/file/work/OntologySummit2013/OntologySummit2013_Communique/OntologySummit2013_Communique_v1-0-0_20130503.pdf. Accessed 08 Feb 2021

21. Abdelghany, A., Darwish, N., Hefni, H.: An agile methodology for ontology development. Int. J. Intell. Eng. Syst. **12**, 170–181 (2019). https://doi.org/10.22266/ijies2019.0430.17

22. Smirnov, A., Levashova, T.: Models of decision support in socio-cyber-physical systems. Inf. Control Syst. 55–70 (2019). https://doi.org/10.31799/1684-8853-2019-3-55-70

23. Rockwell, J., Grosse, I.R., Krishnamurty, S., Wileden, J.C.: A decision support ontology for collaborative decision making in engineering design. In: 2009 International Symposium on Collaborative Technologies and Systems, pp. 1–9. IEEE (2009). https://doi.org/10.1109/CTS.2009.5067456

24. Rockwell, J.A., Grosse, I.R., Krishnamurty, S., Wileden, J.C.: A semantic information model for capturing and communicating design decisions. J. Comput. Inf. Sci. Eng. **10** (2010). https://doi.org/10.1115/1.3462926

25. McGuinness, D.L., van Harmelen, F.: OWL Web Ontology Language Overview. https://www.w3.org/TR/owl-features/. Accessed 08 Feb 2021

26. Protégé. https://protege.stanford.edu/. Accessed 08 Feb 2021

Investigating Oversampling Techniques for Fair Machine Learning Models

Sanja Rančić$^{(\boxtimes)}$, Sandro Radovanović⬚, and Boris Delibašić⬚

Faculty of Organizational Sciences, University of Belgrade, Jove Ilića 154,
11000 Belgrade, Serbia
sr20170133@student.fon.bg.ac.rs

Abstract. Applying machine learning in real-world applications may have various implications on companies, but individuals as well. Besides obtaining lower costs, faster time to decision and higher accuracy of the decision, automation of decisions can lead to unethical and illegal consequences. More specifically, predictions can systematically discriminate against a certain group of people. This comes mainly due to dataset bias. In this paper, we investigate instances oversampling to improve fairness. We tried several strategies and two techniques, namely SMOTE and random oversampling. Besides traditional oversampling techniques, we tried oversampling of instances based on sensitive attributes as well (i.e. gender or race). We demonstrate on real-world datasets (Adult and COMPAS) that oversampling techniques increase fairness, without greater decrease in predictive accuracy. Oversampling improved fairness up to 15% and AUPRC up to 3% with a loss in AUC of 2% .

Keywords: Data preprocessing · Oversampling · SMOTE · Algorithmic fairness · Machine learning

1 Introduction

Usage of algorithmic decision making is increasing over the years. The need for both fast and accurate decisions has overcome expert knowledge. The process of replacing experts in decision making with computer models resulted in the usage of machine learning algorithms in many real-life business applications. For example, descriptive and predictive models are successfully used for client selection in marketing campaigning [17], churn prediction [1], or up-sell and cross-sell models [24]. Even though machine learning models have shown benefits in terms of expenses, accuracy, and speed, one must be aware of the growing issue of fairness in machine learning models. More specifically, machine learning models can and often do, systematically discriminate certain subgroups [4]. To make things worse, this discrimination is based on a personal trait that a person cannot change (or it is not expected to change) such as gender, race, or religion.

Making discrimination that can be observed by inspecting gender, race, or religion in predictions can be a subject of legal consequences [7]. For example, Google's job recommendation engine suggested higher-paid jobs to male individuals, which resulted

© Springer Nature Switzerland AG 2021
U. Jayawickrama et al. (Eds.): ICDSST 2021, LNBIP 414, pp. 110–123, 2021.
https://doi.org/10.1007/978-3-030-73976-8_9

in the European Commission fining Google [9]. Having the above mentioned in mind, one would like to mitigate unwanted discrimination and provide an accurate predictive model as it can be. The problem of unwanted discrimination in the (algorithmic) decision-making process is challenging from a philosophical point of view. Defining the term being fair in the decision-making process is very context-dependent, and the sense of fairness is different based on the utility that one obtains as a result of a predictive model [6, 20]. From a machine learning point of view, one needs to mathematically develop a notion of fairness or use an existing one and integrate it into a machine learning process [27].

In this paper, we focus on data preprocessing techniques for achieving fair predictive models. Although other approaches can be used for fair machine learning models, we selected data preprocessing due to the importance of data preprocessing in the machine learning process. By observing the cross-industry standard process (CRISP-DM) methodology [25] data preparation is one of the most important steps which often take the most time. This paper proposes data preparation techniques for fairness, namely, oversampling for fairness. Oversampling tries to multiply examples or generate new examples in the dataset in such a manner to diminish the effects of discrimination. This is done in such a manner that instances are equalized based on the value of the output attribute and the attribute considered for unwanted discrimination (i.e. gender or race). The proposed technique is evaluated on two datasets known for having a high level of unwanted discrimination. The first one is Adult [16] where female gender instances are discriminated compared to the male gender instances. The other dataset is the COMPAS dataset [10] where Afro-Americans are discriminated. To the best of our knowledge, inspecting various oversampling tactics for fairness has not been done before.

The paper is structured as follows. In Sect. 2 we provide related work. This section covers the notions of fairness in machine learning, as well as existing approaches for achieving fairness. Section 3 provides a methodology of the research. More specifically, we explain the data at hand, proposed methods, and experimental setup. Then, in Sect. 4 we present results and the discussion of the results. Finally, Sect. 5 concludes the paper and provides future research directions.

2   Related Work

Fairness and justice are traditionally discussed in the area of philosophy or social sciences. However, with the development of data privacy and data regulation laws, computer scientists adopted fairness definitions and implemented them in machine learning processes. Having that in mind we first provide notions of fairness in machine learning. Following that, we present existing approaches for dealing with unwanted bias, based on which we derive our methodology.

2.1   Fairness in Machine Learning

Fairness in machine learning considers both group and individual fairness. Individual fairness is a requirement of algorithmic decision-making models that similar individuals receive similar predictions, and thus decisions [23]. Even though this notion seems intuitively satisfied by machine learning models, one can yield individually unfair models [15]. For example, if a predictive model is too complex and overfitting, then a small change in input attributes can result in large differences in outputs. Therefore, individual fairness measures the consistency of predictive models.

Group fairness is most commonly defined as systematic discrimination in algorithmic decision-making tools based on attributes such as race, gender, or religion. Such attributes are called sensitive and they are denoted as s. One individual can belong to either a discriminated group ($s = 1$) or a privileged group ($s = 0$). The discriminated group has a lower overall utility score compared to the privileged group. There are multiple definitions of group fairness. The most prominent one is disparate impact.

$$DI = \frac{E(\hat{y}|s = 1)}{E(\hat{y}|s = 0)} \tag{1}$$

where E presents the mathematical expectation, and $\hat{y}$ the probability of an output attribute value. More specifically, disparate impact can be interpreted as the expected probability of getting a desired outcome and it should be independent of the sensitive attribute. This means that both privileged and discriminated groups should have the same probability of getting a desired outcome. For example, if 60 males and 40 females applied for university enrollment from which 40 males and 20 females were finally enrolled, then DI would be the enrollment ratio percentage per gender. As the percentage of male enrolment is $\frac{40}{60} = 0.667$ and the percentage of female enrolments is $\frac{20}{40} = 0.500$, then $DI = \frac{0.500}{0.667} = 0.75$. Disparate impact can be used both prior to learning a machine learning model and for evaluating the learning process. In the first case, it can be used to evaluate whether a decision-making process was faulty and subject to possible legal consequences. In the latter case, one can inspect if the predictive model is unfair or not [2, 7].

However, one needs referent points to evaluate if a decision-making process is fair. Although there are no formal guidelines about the allowable level of unfairness, one can use the U.S. Equal Employment Opportunity Commission "80% rule" [3]. This rule states that disparate impact should be above 0.8 and lower than 1.25 (1/0.8).

Other group fairness measures, also used in practice, are statistical parity (a linear representation of disparate impact), equality of opportunity [11] (measures whether the expected value for a desired outcome is independent of the sensitive attribute), equality of odds (measures whether expected probability scores should be independent for all possible outcomes of the process) [21].

2.2 Fairness Techniques

Over the past several years several approaches for fairness have been developed. The first naïve approach assumes that a predictive model will be fair as long as the sensitive attribute and the correlated attributes are removed from dataset. This is called fairness through unawareness [6]. Even though it is simple and intuitive, often attributes that can identify the value of the sensitive attribute exist, thus making a predictive model unfair. Due to the unsatisfactory results of the fairness through unawareness approach, other approaches have been developed. All of them can be broadly categorized into three categories.

Pre-processing techniques are developed to reduce or remove the discrimination in the data. One can alter the dataset by selecting the attributes, or selecting the instances, but also changing the values in the data. One intuitive approach is to remove the attributes that are correlated with the sensitive attribute [14]. This approach yields better fairness metrics, but due to the removal of attributes, the predictive performance is much lower. Also, the removal of attributes that are correlated with the sensitive attribute does not guarantee a fair predictive model. Attribute interactions can indicate unfairness but are not observed if the correlation coefficient is used. One can also massage the output attribute [13]. Massaging the output attribute changes the output value of some instances from the discriminated group to be positive, and also for some instances from the privileged group to be positive. Although fairness is improved, changing data distribution results in lower data accuracy and this lower predictive accuracy. Another approach is assigning weights to the instances. One algorithm called reweighing [14] assigns weights to each instance using a chi-square test like metric. More specifically, weight of an instance is obtained as a ratio of expected and observed number of instances with a specific value of the sensitive attribute and specific outcome value. Another approach that is recently applied is oversampling of the discriminated group [22, 26]. Fair class balancing attempts to increase the importance of the discriminated group by generating new examples that belong to the discriminated group.

In-processing techniques attempt to alter the learning algorithm and introduce fairness during the model learning phase. This set of techniques are considered to be the most complex since one needs to define a loss function, a regularization function, or constraints for the learning algorithm and provide an optimization procedure to solve the given problem. One can find examples where statistical parity is introduced as a constraint in the logistic regression algorithm [27]. Similarly, one can find constraints for equalized odds in logistic regression [21]. However, instead of using constraints one can use regularization terms and hyper-parameters to control for unfairness [12]. Finally, one can use adversarial learning [18] where one learns to predict the output attribute while not being able to predict the sensitive attribute.

Finally, one can attempt to provide fair predictive models using *post-processing* *techniques*. Techniques from this group adjust the probability scores to obtain fair predictions. For example, one can analytically select the best possible decision threshold for which the best possible fairness is achieved [19]. This can be also presented as a linear optimization problem with the goal function of satisfaction of statistical parity, equal opportunity, or equalized odds by changing the decision threshold given the accuracy constraints. Another approach introduces linear programming for selecting the best trade-off between prediction accuracy and fairness, where a decision-maker can provide constraints regarding both accuracy and fairness [11].

Based on the related work in the area of fairness in machine learning, we believe that data pre-processing is appropriate for achieving fairness. This belief is based on the principle of *unfairness-in unfairness-out*. Thus, by adjusting the data to be fair, one would be able to obtain better predictive models in terms of fairness. More specifically, if one does not clean the dataset and remove unwanted bias, then the resulting prediction model will be unfair. The approach that we propose is not intrusive, thus does not change the values of the data points themselves, but signals the learning algorithm of unfair instances. Therefore, we propose the usage of oversampling techniques that will yield a higher representation of the instances of the discriminated group. Even though oversampling has been done in various machine learning applications, effect of oversampling on fairness is not explored enough.

3 Methodology

The main idea of this paper is to obtain fair models that are approximately as accurate as their discriminatory pairs by oversampling the data. In this section, we will propose two approaches to achieving that, as well as present the data that was used for experiments and the experimental setup.

3.1 Oversampling for Fairness

While observing the data, in most cases, we can conclude that instances that belong to the unprivileged group and at the same time have the favorable target value represent a smaller percentage.

Oversampling is a procedure of adjusting the outcome class distribution in the data at hand so that the ratio of the outcomes is equalized. This is done for the class imbalance problems where the learning algorithm has the problem of identification of the outcome class with fewer instances. In those cases, one either replicate the existing instances of the class with fewer instances or generate new instances. More specifically, the representation of class outcome in data poses a problem that can be solved using the oversampling of instances. The usage of oversampling has shown interesting results in practical settings, i.e. it often improves the predictive performance of models [5].

To equate the privileged and unprivileged group by the outcome, we evaluate four oversampling strategies:

1) *Traditional oversampling.* Oversampling intended for equalizing the number of instances in output classes. Here oversampling is performed such that the number of instances of the desired outcome ($y = 1$) is equal to the number of instances of the undesired outcome ($y = 0$). This type of oversampling does not take into account a sensitive attribute.

2) *Oversampling on a sensitive attribute.* Since the problem of fairness can be observed as a problem of underrepresentation of the discriminated group in the dataset, we would like to inspect what will happen if one increases the number of instances from the discriminated group. Therefore, we perform oversampling based on the value of the sensitive attribute. This way our dataset will have exactly the same number of instances from both discriminated and privileged groups.

3) *Equalized number of discriminated group instances.* In this approach, we generate only instances from the discriminated group. More specifically, we equalize the number of examples with the desired outcome ($y = 1$) with the number of examples with the undesired outcome ($y = 0$), but only for the discriminated group ($s = 1$).

This way, one will achieve an approximately equal average value of the output for the discriminated group. In this case, the assumption is that the problem is solely in the instances of the discriminated group. By increasing the number of instances with the desired outcome from the discriminated group better representation of the discriminated group will be made.

4) *Equalized desired outcome for both discriminated and privileged groups.* This approach oversamples both privileged ($s = 1$) and unprivileged groups ($s = 0$). Here the dataset is divided into two distinct datasets. One regarding discriminated instances, and another regarding privileged instances. Then, independently we equalize the number of instances based on the outcome attribute. This will result in unequal datasets due to the initial distribution of the data. However, by doing this one will achieve an approximately equal relative number of the desired outcome across the sensitive attribute, i.e. the same percentage of the instances with the desired outcome in both discriminated and privileged groups.

Oversampling techniques that will be used are Random Oversampling and SMOTE [5]. Random Oversampling performs sampling with replacement of the minority group. This procedure generates a new dataset that contains an increased number of instances compared to the original one by appending the same instances from the original dataset. The effects of Random Oversampling is presented in Fig. 1. Instances that have more weighted outlines are instances that are oversampled. Due to the random procedure, one instance is replicated multiple times. Although strategy *Traditional oversampling* seems to result in the same dataset as *Equalized desired outcome for both discriminated and privileged groups* there is one crucial difference. Traditional oversampling is unaware of the sensitive attribute s, thus the number of instances with $y = 1$ will be the same as the number of instances having $y = 0$. This may result in an unequal number of examples based on the sensitive attribute s. In the latter strategy, the dataset is divided by s, and then the number of instances based on y are equalized. More specifically, if one takes only $s = 1$ (or $s = 0$) instances, then the number of instances with a desired ($y = 1$) and undesired ($y = 0$) outcome will be the same.

SMOTE algorithm [5] generates synthetic instances by observing the local neighborhood of randomly selected instances. This is done by observing k nearest neighbor instances of the selected instance and generating new instances on the line segments between the selected instance and k nearest neighbor instances for each combination of features. Then, for each feature and randomly selected line segment (out of k) random value on a line segment is selected. Once every feature is constructed a synthetic instance is created. To the best of our knowledge, inspecting various oversampling tactics for fairness has not been done before. The process of oversampling instances with SMOTE with the proposed strategies is presented in Fig. 2. New instances will be generated on the lines (i.e. linear combination of input attributes of similar instances) presented on the figures.

Similarly, as in Random oversampling, the difference between *Traditional oversampling* and *Equalized desired outcome for both discriminated and privileged groups* is that traditional oversampling will generate new instances as a linear combination of instances having the desired outcome ($y = 1$) regardless of the sensitive attribute, while the latter approach differentiates new instances based on the sensitive attribute.

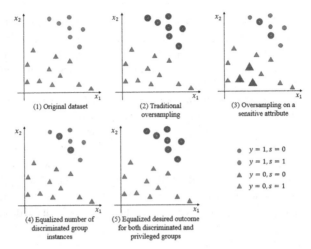

Fig. 1. Random oversampling with proposed oversampling strategies

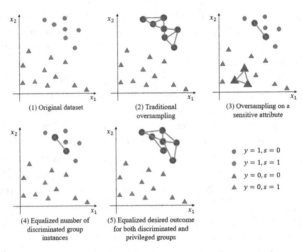

Fig. 2. SMOTE with proposed oversampling strategies

3.2 Data

The experiments are performed on two datasets well known for having unfair outcomes: Adult and COMPAS. Adult [16] is a dataset that involves predicting personal income levels as above or below $50,000 per year based on personal details (e.g., relationship, education level, etc.). It is believed that the female gender is discriminated compared to the male gender. Therefore, one would like to reduce the gender gap and create a prediction model that does not create gender inequality. COMPAS [10] is a dataset used for predicting whether a perpetrator will repeat the crime or not based on personal details and crime details. Due to various reasons (i.e. Afro-Americans commit more crimes), predictive systems create discriminative predictions toward Afro-Americans, thus sentencing them more often (compared to Caucasians). Both datasets have a high class imbalance and high fairness imbalance.

3.3 Experimental Setup

Two experiments were performed. As a learning algorithm, we utilized Naïve Bayes (NB), Logistic Regression (LR), and Random Forest (RF). The Naïve Bayes algorithm is one of the simplest algorithms in the area of machine learning. It is selected to test whether fairness is related to the complexity of the algorithms [21]. We used value one for Laplace smoothing in all experiments. The logistic regression algorithm presents a linear classifier that models the probability of an event based on a logit model. In this research, we used a modification of the logistic regression algorithm called Ridge logistic regression which is a modification that uses regularization of the attribute coefficients using the coefficient L_2 unit ball. Finally, we used the Random Forest algorithm. This algorithm uses an ensemble of 100 decision trees. Each decision tree uses a random 10% of attributes and a random 70% of instances.

We used the same metrics for both the model accuracy and the fairness. To evaluate the predictive performance we utilize the area under the curve (AUC) and the area under the precision-recall curve (AUPRC). AUC can be interpreted as a probability that a random instance with a desired outcome has a higher probability from than a random instance with an undesired outcome [8]. AUPRC calculates the precision and recall of the predictive model for each value of the decision threshold and calculates the area constructed by the precision-recall curve. AUPRC is recommended for class imbalanced datasets [8]. For both AUC and AUPRC the best possible value is one. Values lower than one indicate lower predictive performance of predictive models.

As a measure of fairness, we utilize disparate impact (DI) as presented in Eq. (1). This measure is bounded between zero and infinity, where a value equal to one indicates perfect fairness. Also, this measure is not symmetrical, meaning that values 0.5 and 2 present the same level of unfairness. To ensure that the best possible value is one we calculate DI as $\min(DI, DI^{-1})$. This way the maximum value is one and this value present perfect fairness.

Experiments are conducted using ten-fold cross-validation on both datasets. We report average results as well as standard deviation obtained on the test set. Oversampling techniques are conducted only on the training set while leaving the test set intact. This way we ensure a fair comparison of the results. As a baseline approach, we use the no pre-processing approach. For experimenting, we performed the four above-mentioned oversampling techniques. Those are traditional oversampling (denoted as *Oversampling on y*), oversampling on the sensitive attribute (denoted as *Oversampling on s*), equalized number of discriminated group instances (denoted as *Equalized s = 1*), and equalized the desired outcome for both discriminated and privileged groups (denoted as *Equalized y for both s*).

4 Results and Discussion

The results and discussion section are divided into two separate parts. The first part discusses the results on the Adult dataset, where we can compare the results between the proposed methods and the baseline approaches. Since three measures are used, two for predictive accuracy and one for fairness, there are three sets of results. Values of AUC are presented in Table 1.

Table 1. Results for Adult dataset – AUC

	Algo.	No preprocessing	Oversampling on y	Oversampling on s	Equalized $s = 1$	Equalized y for both s
SMOTE	NB	**0.862 ± 0.010**	0.834 ± 0.010	0.858 ± 0.009	0.850 ± 0.010	0.827 ± 0.010
	LG	**0.901 ± 0.006**	0.868 ± 0.007	**0.901 ± 0.006**	0.876 ± 0.007	0.865 ± 0.007
	RF	**0.901 ± 0.007**	0.889 ± 0.009	**0.901 ± 0.007**	0.893 ± 0.009	0.888 ± 0.008
RO	NB	**0.862 ± 0.010**	0.862 ± 0.010	0.858 ± 0.010	0.864 ± 0.011	0.850 ± 0.010
	LG	**0.901 ± 0.006**	**0.901 ± 0.006**	**0.901 ± 0.006**	**0.901 ± 0.006**	0.899 ± 0.006
	RF	**0.901 ± 0.007**	0.899 ± 0.008	0.900 ± 0.007	0.900 ± 0.008	0.899 ± 0.008

The best performing model for each algorithm is presented in bold letters. This level of predictive performance indicates a superb predictive model that can differentiate between output values. As it can be observed from the table, AUC did not improve by using oversampling techniques. However, the majority of them remain at a similar or even the same level as without any intervention in the dataset. For example, for the Naïve Bayes algorithm, the biggest decrease in the performance is for Equalized y for both s with SMOTE oversampling where AUC decreased 3.5%. A similar drop in performance can be observed for logistic regression, while random forest had a stable AUC regardless of the oversampling technique and oversampling tactic.

The results for AUPRC are presented in Table 2. The best performance is presented in bold letters. More specifically, the best performing predictive model is the random forest model with AUPRC equal to 0.775. Based on this value we can say that obtained model is almost twice as good compared to the random model.

Table 2. Results for Adult dataset – AUPRC

	Algo.	No preprocessing	Oversampling on y	Oversampling on s	Equalized $s = 1$	Equalized y for both s
SMOTE	NB	**0.704 ± 0.014**	0.685 ± 0.013	0.692 ± 0.015	0.700 ± 0.014	0.679 ± 0.013
	LG	**0.758 ± 0.011**	0.684 ± 0.013	**0.758 ± 0.011**	0.704 ± 0.012	0.678 ± 0.012
	RF	**0.775 ± 0.015**	0.742 ± 0.020	0.772 ± 0.015	0.753 ± 0.018	0.742 ± 0.019
RO	NB	0.704 ± 0.014	0.704 ± 0.014	0.696 ± 0.016	**0.709 ± 0.015**	0.690 ± 0.013
	LG	**0.758 ± 0.011**	0.756 ± 0.011	0.757 ± 0.012	0.755 ± 0.011	0.755 ± 0.013
	RF	**0.775 ± 0.015**	0.758 ± 0.017	0.772 ± 0.015	0.763 ± 0.017	0.763 ± 0.016

In Table 3 we present fairness metrics (DI) obtained using both SMOTE and RO, as well as in the no preprocessing setting. One can observe that fairness improved in almost all oversampling strategies. However, the oversampling technique that provided the best fairness result is the equalized outcome (y) for both discriminated and privileged groups. More specifically, one can notice the increase of DI with at least 0.5%, up to even 16%. An increase in fairness is much higher compared to the no preprocessing models. Several conclusions can be made. Although it seems like an intuitive approach oversampling the sensitive attribute, fairness decreased. The reason why the fairness is lower compared to the no preprocessing model is the fact that discrimination is not only observed in the number of examples belonging to the discriminated group but in their outcomes as well. Therefore, by generating instances from the discriminated group we generated more instances with an undesired outcome, thus amplifying unwanted discrimination. Other approaches equalized the desired outcome, thus increased fairness.

Table 3. Results for Adult dataset – DI

	Algo.	No preprocessing	Oversampling on y	Oversampling on s	Equalized $s = 1$	Equalized y for both s
SMOTE	NB	0.421 ± 0.088	0.338 ± 0.060	$\mathbf{0.450 \pm 0.066}$	0.370 ± 0.069	0.426 ± 0.085
	LG	0.389 ± 0.023	0.433 ± 0.022	0.386 ± 0.022	0.419 ± 0.021	$\mathbf{0.464 \pm 0.022}$
	RF	0.393 ± 0.023	0.423 ± 0.018	0.388 ± 0.023	0.408 ± 0.019	$\mathbf{0.439 \pm 0.021}$
RO	NB	0.421 ± 0.088	0.429 ± 0.100	0.427 ± 0.096	0.391 ± 0.096	$\mathbf{0.576 \pm 0.128}$
	LG	0.389 ± 0.023	0.462 ± 0.023	0.375 ± 0.023	0.401 ± 0.021	$\mathbf{0.552 \pm 0.023}$
	RF	0.393 ± 0.023	0.405 ± 0.021	0.388 ± 0.022	0.387 ± 0.021	$\mathbf{0.414 \pm 0.020}$

Next, we present the results for the COMPAS dataset. The predictive performance of the COMPAS dataset is presented in Table 4. As one can observe, there is no clear winner. More specifically, predictive performance in terms of AUC is similar for every tested approach.

Table 4. Results for COMPAS dataset – AUC

	Algo.	No preprocessing	Oversampling on y	Oversampling on s	Equalized $s = 1$	Equalized y for both s
SMOTE	NB	0.722 ± 0.013	$\mathbf{0.725 \pm 0.014}$	0.723 ± 0.012	0.724 ± 0.014	0.724 ± 0.013
	LG	$\mathbf{0.739 \pm 0.010}$	$\mathbf{0.739 \pm 0.011}$	0.738 ± 0.009	0.738 ± 0.011	0.738 ± 0.011
	RF	$\mathbf{0.698 \pm 0.015}$	0.692 ± 0.015	0.696 ± 0.014	0.690 ± 0.017	0.694 ± 0.016
RO	NB	$\mathbf{0.722 \pm 0.013}$	$\mathbf{0.722 \pm 0.013}$	$\mathbf{0.722 \pm 0.014}$	$\mathbf{0.722 \pm 0.014}$	$\mathbf{0.722 \pm 0.014}$
	LG	$\mathbf{0.739 \pm 0.010}$	$\mathbf{0.739 \pm 0.010}$	$\mathbf{0.739 \pm 0.010}$	$\mathbf{0.739 \pm 0.010}$	$\mathbf{0.739 \pm 0.010}$
	RF	$\mathbf{0.698 \pm 0.015}$	0.695 ± 0.015	0.692 ± 0.014	0.694 ± 0.016	0.694 ± 0.014

A similar conclusion can be made for AUPRC. The results are presented in Table 5. The results suggest that models using oversampling techniques and no preprocessing model have similar predictive performance in terms of AUPRC.

Table 5. Results for COMPAS dataset – AUPRC

	Algo.	No preprocessing	Oversampling on y	Oversampling on s	Equalized $s = 1$	Equalized y for both s
SMOTE	NB	0.665 ± 0.023	**0.666 ± 0.024**	0.664 ± 0.022	0.666 ± 0.024	**0.666 ± 0.024**
	LG	0.693 ± 0.019	0.693 ± 0.021	**0.694 ± 0.019**	0.692 ± 0.020	0.692 ± 0.020
	RF	**0.634 ± 0.028**	0.628 ± 0.026	0.626 ± 0.028	0.622 ± 0.030	0.625 ± 0.028
RO	NB	**0.665 ± 0.023**	0.664 ± 0.022	0.664 ± 0.023	**0.665 ± 0.023**	**0.665 ± 0.023**
	LG	0.693 ± 0.019	0.693 ± 0.019	0.693 ± 0.019	0.693 ± 0.020	**0.694 ± 0.020**
	RF	**0.634 ± 0.028**	0.630 ± 0.027	0.622 ± 0.029	0.627 ± 0.027	0.627 ± 0.021

Finally, we present fairness in terms of DI in Table 6. In this setup, equalizing the desired outcome for both discriminated and privileged groups performed the best. However, an increase in fairness can be observed in results from every oversampling technique.

Table 6. Results for COMPAS dataset – DI

	Algo.	No preprocessing	Oversampling on y	Oversampling on s	Equalized $s = 1$	Equalized y for both s
SMOTE	NB	0.593 ± 0.055	0.629 ± 0.052	0.571 ± 0.059	0.636 ± 0.053	**0.645 ± 0.054**
	LG	0.804 ± 0.032	0.823 ± 0.029	0.781 ± 0.035	0.832 ± 0.029	**0.834 ± 0.029**
	RF	0.807 ± 0.020	0.824 ± 0.023	0.792 ± 0.033	**0.830 ± 0.026**	0.827 ± 0.027
RO	NB	0.593 ± 0.055	0.610 ± 0.047	0.600 ± 0.055	0.626 ± 0.053	**0.629 ± 0.054**
	LG	0.804 ± 0.032	0.819 ± 0.029	0.807 ± 0.032	0.827 ± 0.029	**0.829 ± 0.028**
	RF	0.807 ± 0.020	0.818 ± 0.019	0.811 ± 0.027	**0.823 ± 0.022**	0.821 ± 0.025

To summarize, oversampling influences predictive models. It is shown in many applications that oversampling can improve predictive performance [5]. However, in this experiment values of AUC and AUPRC did not improve. Predictive performance remained approximately the same as if no oversampling was conducted. On the other side, fairness did improve. Using equalizing the desired outcome for both discriminated and privileged groups achieved the best performance. This is expected due to the equalization of the number of the desired outcome in both discriminated and privileged groups. To some extent the approach where one equalizes the number of discriminated group instances performs well too regarding fairness. Naïve Bayes algorithm decreased fairness. It is worth to notice that usage of oversampling based on the sensitive attribute is not a tactic one should seek. Although it seems intuitive, the resulting dataset generates greater unwanted discrimination due to the generation of instances that are of undesired outcome and inherited bias that discriminated group has a lower percentage of achieving a desired outcome and thus unfairness is amplified. Traditional oversampling often performed satisfactorily. This means that predictive performance remained the same, while fairness increased.

Based on the results, the authors suggest trying the *Equalized desired outcome for both discriminated and privileged groups* strategy first. For both Adult and COMPAS datasets, this strategy produced the best fairness conditions in most of the cases. Additionally, predictive performance was shown to be on a satisfactory level. Strategies *Equalized number of discriminated group instances* and *Oversampling on a sensitive attribute* also performed well on both datasets. As for the oversampling technique to be used, it is recommended to use SMOTE, rather than Random Oversampling.

5 Conclusions

This paper deals with the effect of oversampling techniques and tactics on algorithmic decision-making fairness. We experimented and evaluated two oversampling techniques, namely SMOTE and random oversampling, using four oversampling techniques. More specifically, we evaluated traditional oversampling, oversampling on the sensitive attribute, equalized the number of discriminated group instances, and equalized the desired outcome for both discriminated and privileged groups. The experiments are conducted on two datasets known for unwanted discrimination. Namely, Adult and COMPAS. Additionally, to get a better insight we performed experiments using three algorithms – Naïve Bayes, Logistic Regression, and Random Forest.

The results suggest that oversampling managed to improve fairness of the predictive model without hurting prediction accuracy. More specifically, on the Adult dataset fairness increased several percent, up to even 15%, while AUC decreased up to 3% and AUPRC up to 10%. A similar conclusion can be drawn on the COMPAS dataset where fairness increased 3%, up to 5%, with AUC and AUPRC approximately the same.

As a part of future work, we want to develop a preprocessing pipeline that will identify unfairness in the data, cleanse it, and provide a fair dataset to the learning algorithm. This pipeline should consist of attribute selection (or weighting), instance selection (or weighting), and data transformation. The core idea is to provide a guarantee that data are fair before learning a predictive model.

Acknowledgments. This work was partially funded in part by the ONR/ONR Global under Grant N62909-19-1-2008. We would like to thank Saga New Frontier Group for supporting this research.

References

1. Ahmad, A.K., Jafar, A., Aljoumaa, K.: Customer churn prediction in telecom using machine learning in big data platform. J. Big Data **6**(1), 1–24 (2019). https://doi.org/10.1186/s40537-019-0191-6
2. Barocas, S., Selbst, A.D.: Big data's disparate impact. Calif. Law Rev. **104**, 671 (2016)
3. Biddle, D.A.: Adverse Impact and Test Validation: A Practitioner's Handbook. Infinity Publishing (2012)

4. Binns, R.: Fairness in machine learning: lessons from political philosophy. In: Conference on Fairness, Accountability and Transparency, pp. 149–159. PMLR, January 2018

5. Chawla, N.V., Bowyer, K.W., Hall, L.O., Kegelmeyer, W.P.: SMOTE: synthetic minority over-sampling technique. J. Artif. Intell. Res. **16**, 321–357 (2002)

6. Chen, J., Kallus, N., Mao, X., Svacha, G., Udell, M.: Fairness under unawareness: assessing disparity when protected class is unobserved. In: Proceedings of the Conference on Fairness, Accountability, and Transparency, pp. 339–348, January 2019

7. Corbett-Davies, S., Pierson, E., Feller, A., Goel, S., Huq, A.: Algorithmic decision making and the cost of fairness. In: Proceedings of the 23rd ACM SigKDD International Conference on Knowledge Discovery and Data Mining, pp. 797–806, August 2017

8. Cortes, C., Mohri, M.: AUC optimization vs. error rate minimization. Adv. Neural. Inf. Process. Syst. **16**, 313–320 (2003)

9. Datta, A., Tschantz, M.C., Datta, A.: Automated experiments on ad privacy settings: a tale of opacity, choice, and discrimination. Proc. Privacy Enhancing Technol. **2015**(1), 92–112 (2015)

10. Dressel, J., Farid, H.: The accuracy, fairness, and limits of predicting recidivism. Sci. Adv. **4**(1), eaao5580 (2018)

11. Hardt, M., Price, E., Srebro, N.: Equality of opportunity in supervised learning. In: Advances in Neural Information Processing Systems, pp. 3315–3323 (2016)

12. Horesh, Y., Haas, N., Mishraky, E., Resheff, Y., Meir Lador, S.: Paired-consistency: an example-based model-agnostic approach to fairness regularization in machine learning. In: Cellier, Peggy, Driessens, Kurt (eds.) ECML PKDD 2019. CCIS, vol. 1167, pp. 590–604. Springer, Cham (2020). https://doi.org/10.1007/978-3-030-43823-4_47

13. Kamiran, F., Calders, T.: Classifying without discriminating. In: 2009 2nd International Conference on Computer, Control and Communication, pp. 1–6. IEEE, February 2009

14. Kamiran, F., Calders, T.: Data preprocessing techniques for classification without discrimination. Knowl. Inf. Syst. **33**(1), 1–33 (2012)

15. Kleinberg, J., Ludwig, J., Mullainathan, S., Rambachan, A.: Algorithmic fairness. In: AEA Papers and Proceedings, vol. 108, pp. 22–27, May 2018

16. Kohavi, R.: Scaling up the accuracy of Naive-Bayes classifiers: a decision-tree hybrid. In: Proceedings of the Knowledge Discovery in Data, vol. 96, pp. 202–207, August 1996

17. Ładyżyński, P., Żbikowski, K., Gawrysiak, P.: Direct marketing campaigns in retail banking with the use of deep learning and random forests. Expert Syst. Appl. **134**, 28–35 (2019)

18. Petrović, A., Nikolić, M., Radovanović, S., Delibašić, B., Jovanović, M.: FAIR: Fair Adversarial Instance Re-weighting. arXiv preprint arXiv:2011.07495 (2020)

19. Pleiss, G., Raghavan, M., Wu, F., Kleinberg, J., Weinberger, K.Q.: On fairness and calibration. In: Advances in Neural Information Processing Systems, pp. 5680–5689 (2017)

20. Radovanović, S., Petrović, A., Delibašić, B., Suknović, M.: Making hospital readmission classifier fair–what is the cost? In: Central European Conference on Information and Intelligent Systems, pp. 325–331. Faculty of Organization and Informatics Varazdin (2019)

21. Radovanović, S., Petrović, A., Delibašić, B., Suknović, M.: Enforcing fairness in logistic regression algorithm. In: 2020 International Conference on INnovations in Intelligent SysTems and Applications (INISTA), pp. 1–7. IEEE, August 2020

22. Rančić, S., Radovanović, S., Suknović, M.: Effects of data preprocessing for fairness in machine learning. In: Proceedings of the XVII International Symposium – SymOrg, Zlatibor, Serbia, 7–10 September 2020, pp. 255–262 (2020)

23. Sharifi-Malvajerdi, S., Kearns, M., Roth, A.: Average individual fairness: algorithms, generalization and experiments. In: Advances in Neural Information Processing Systems, pp. 8242–8251 (2019)

24. Syam, N., Sharma, A.: Waiting for a sales renaissance in the fourth industrial revolution: machine learning and artificial intelligence in sales research and practice. Ind. Mark. Manag. **69**, 135–146 (2018)
25. Wirth, R., Hipp, J.: CRISP-DM: towards a standard process model for data mining. In: Proceedings of the 4th International Conference on the Practical Applications of Knowledge Discovery and Data Mining, pp. 29–39. Springer, London, April 2000
26. Yan, S., Kao, H.T., Ferrara, E.: Fair class balancing: enhancing model fairness without observing sensitive attributes. In: Proceedings of the 29th ACM International Conference on Information and Knowledge Management, pp. 1715–1724, October 2020
27. Zafar, M.B., Valera, I., Gomez-Rodriguez, M., Gummadi, K.P.: Fairness constraints: a flexible approach for fair classification. J. Mach. Learn. Res. **20**(75), 1–42 (2019)

Using AI to Advance Factory Planning: A Case Study to Identify Success Factors of Implementing an AI-Based Demand Planning Solution

Ulrike Dowie and Ralph Grothmann[✉]

Digital Industries, Siemens AG, Munich, Germany
Ralph.Grothmann@siemens.com

Abstract. Rational planning decisions are based upon forecasts. Precise forecasting has therefore a central role in business. The prediction of customer demand is a prime example. This paper introduces recurrent neural networks to model customer demand and combine the forecast with uncertainty measures to derive decision support of the demand planning department. It identifies and describes the keys to the successful implementation of an AI-based solution: bringing together data with business knowledge, AI methods and user experience, and applying agile software development practices.

Keywords: Agile software development · AI project success factors · Deep learning · Demand forecasting · Forecast uncertainty · Neural networks · Supply chain management

1 Introduction

Forecasting customer demand is probably one of the most challenging tasks in managing a supply chain. Accurate sales perspectives allow decreasing inventories while customer satisfaction is increased due to higher delivery reliability and capability. New directions in forecasting customer demand come from the field of artificial intelligence (AI). Deep learning technologies can identify demand patterns from historical data that result from seasonality (e.g. different characteristics of quarters), internal planning parameters (e.g. product lifecycle) or external influences (e.g. macro-economic factors). These patterns can be exploited to make predictions of the customer demand in short- and mid-term (up to 24 months) related to the existing data history. Confidence intervals can be derived to handle the forecast uncertainty.

In addition to this uncertainty, the planning process for material ordering in many organizations is an iterative procedure. Depending on the frequency of material orders, the number of organization units involved, and the specific software infrastructure used, this process requires considerable manual effort. Therefore, replacing human judgment as planning input with historical data and automating this part of the planning procedure is a promising way to optimize material management.

© Springer Nature Switzerland AG 2021
U. Jayawickrama et al. (Eds.): ICDSST 2021, LNBIP 414, pp. 124–134, 2021.
https://doi.org/10.1007/978-3-030-73976-8_10

How to do this, i.e. how to apply AI to make planning input available in an automated fashion, in a cost-effective way, is the subsequent question.

In this paper a case study of developing a demand planning solution is described that is based on recurrent neural networks (RNN) and uncertainty evaluation based on the related residual errors of the model. Section 2 details the research design. Section 3 introduces the specific kind of AI used, namely RNN, and their application in demand forecasting. Section 4 addresses ways of measuring forecast uncertainty. Section 5 gives empirical results concerning the benefits of the solution for the planning process and its costs, and Sect. 6 outlines the identified success factors for implementing AI projects. Section 7 summarizes the primary findings and points to trends in AI solutions.

2 Research Design

The case study at hand served to answer the following research questions:

1. How can Neural Networks be applied to forecast demand, and what needs to be done to take uncertainty into account?
2. What are the success factors of implementing an AI-based demand planning solution?

This study had an exploratory character, and by answering these questions, it was intended to provide a set of recommendations for further implementations of AI projects.

The unit of analysis is a software development project which aimed to use AI to improve demand forecast accuracy, automate demand planning and thus replace part of the planning workforce.

"Successful implementation" was thus operationalized by reaching these aims of the project. Another criterion of project success was seeing continued use of the developed demand planning solution.

The data used in this case study were interviews with project members and stakeholders and frequent user feedback from material planners, both during development of the demand planning solution and after it was set productive.

3 Forecasting with Neural Networks

Modern neuro-informatics models, like time-delay recurrent neural networks (RNN), offer significant benefits for dealing with the typical challenges associated with forecasting. With their universal approximation properties, neural networks (NN) make it possible to describe non-linear relationships between a large number of external factors and at least one (or many) target variables [3]. In contrast, conventional econometrics generally use linear models (e.g. autoregressive models (AR), multivariate linear regression) which, for all that they facilitate efficient calculation of links, provide only inadequate models for non-linear dynamics. Other conventional time series analysis procedures (such as ARMA, ARIMA, ARMAX) remain confined to linear systems [9].

Probably the widest imaginable range of models is discussed within the class of neural networks [3]. For example, in terms of the data flow in the model, it is possible to draw a distinction between "feedforward" and (time) recurrent neural networks. In a feedforward neural network (FNN), data from an upstream layer is propagated to downstream layers only. Furthermore, there is no provision for any links between the neurons in a single layer. In contrast, (time-delay) recurrent neural networks (RNN) include links which transfer data from a downstream layer (time-delayed) to upstream layers [3, 5]. It is noteworthy that any equation for a neural network can be portrayed in graphic form by means of an architecture which represents the individual layers of the network in the form of nodes and the links between the layers in the form of borders. This relationship is known as correspondence principle between equations, architectures and the local algorithms associated with them [5]. For example, the "error back propagation algorithm" needs only locally available data from the forward and reverse flow of the network in order to calculate the partial formulations of the error function of the neural network according to the weights of a given layer during training [3]. The use of local algorithms here provides an elegant basis for the expansion of the neural network with a view to the modeling of high complexity (large) systems. Used in combination with an appropriate (stochastic) learning rule, it is possible to use the gradients as a basis for the identification of robust minima for network error function [5].

In the case described, RNN were used as discussed in [5] to model and forecast the customer demand in dependence of seasonal effects and macro-economic data as an open dynamic system. The time characteristic for the system is described partially by means of an autonomous (sub-)dynamic and partially by means of external variables. This modeling framework is comparable with a (non-linear) regression approach. The technique of finite unfolding in time is used [3]. The underlying idea here is that any RNN can be reformulated to form an equivalent feedforward neural network using so-called "shared weights". The actual training on the network can then be conducted using the "error-back propagation-through-time" and an appropriate (stochastic) learning rule for the weight update [5]. An advantage of the RNN is the moderate usage of free parameters. In an FNN an expansion of the time delayed input information increases automatically the number of weights, whereas in the RNN the shared matrices are reused if more delayed input information is needed. Consequently, potential over-fitting is not so dangerous as in the training of FNN. In other words: due to the inclusion of the temporal structure into the network architecture, our approach is applicable to tasks where only a small training set is available. This is often the case in customer demand forecasting due to e.g. short product life cycles and fast changing markets.

For further details on the RNN and more sophisticated model architectures used see [5].

4 Planning Under Uncertainty

The results of the neural networks are predictions of the quantities - demand or sales figures - for the products under consideration. The forecasts are the basis for the demand planning, i.e. the expected sales figures determine the material, production and logistics workflow in the factory. However, the forecast usually cannot be used directly to create the demand plan, since the forecast is not perfect, i.e. it incorporates uncertainty.

The experiences gained during the financial crisis or Corona virus pandemic have triggered a far-reaching discussion on the limitations of quantitative forecasting models and made planners very conscious of risk [1]. In order to understand risk distributions, traditional risk management uses diffusion models. Risk is understood as a random walk, in which the diffusion process is calibrated by the observed past error of the underlying model [2]. The starting point is the analysis of model residual errors resp. error distribution. The moments of the distribution of the residual errors is used to calibrate the random walk models, i.e. the assumed diffusion process. The error distribution is used to derive a risk measure that is considered when the forecasts are transformed into a demand plan.

It is important to note that in demand forecasting the left- and the right-hand side of the error distribution should not be treated equally. This means it does make a difference for the planning proposal if demand is over- or underestimated. Overestimations will lead to stock inventories, while underestimates may cause supply shortage, i.e. the customer demand cannot be satisfied. Under the assumption that increased inventory levels do not hurt the company as much as the shortness of supplies, focus is set on those parts of the error distribution where the forecast is smaller than the observed demand. Tracking back the source for large errors in the error distribution is also valuable to study the effect of e.g. black swans that can hurt the demand planning. The error distribution is typically estimated from test data or the most recent forecasts for which observed data is available.

In contrast to this approach uncertainty and risk can also be derived from ensemble forecasts in order to provide important insights into complex risk relationships [4], since internal (unobserved) model variables can be reconstructed from the trend in observed variables (observables). If the system identification (i.e. learning of the demand/sales dynamics) is calculated repeatedly for RNN models, an ensemble of solutions will be produced, which all have a forecast error of zero in the past, but which differ from one another in the future. Since every model gives a perfect description of the observed data, the complete ensemble is the true solution. A way to simplify the forecast is to take the average of the individual ensemble members as the expected value, provided the ensemble histogram is unimodal in every time step.

In addition to the expected value, the bandwidth of the ensemble is considered, i.e. its distribution. The form of the ensemble is governed by differences in the reconstruction of the hidden system variables from the observables: for every finite volume of observations there is an infinite number of explanation models which describe the data perfectly, but differ in their forecasts, since the observations make it possible to reconstruct the hidden variables in different forms during the training. In other words, our risk concept is based on the partial observability of the world, leading to different reconstructions of the hidden variables and thus, different future scenarios. Since all scenarios are perfectly consistent with the history, it is unknown which of the scenarios describes the future demand figures best and risk emerges [4, 5].

In developing the demand planning solution, the concept of analyzing residual error distributions is applied to measure the uncertainty of the demand forecasts for each product individually. This means, that the forecast is adjusted by a measure derived from the tail of the error distribution where the demand is underestimated, i.e. the demand prediction is smaller than the actual customer demand. In addition, prior knowledge is incorporated in the planning proposal. The sources of prior knowledge are orders at hand, information from business development and sales.

5 Results of the Demand Forecasting Project: Benefits and Costs

5.1 Benefits

Measuring benefits in AI projects is challenging, particularly when it comes to risk reduction [10]. For demand forecasting projects, benefits cover more than risk reduction, and KPIs become available that can be quantified rather than estimated.

As a result of the implemented demand planning solution, which provides weekly demand forecasts for material ordering, material stock is reduced by 5% on average. In addition, costs for short-term express orders of material have been avoided repeatedly. However, it's important to note that these stock and cost reductions cannot be attributed to the demand forecasts in a mono-causal way. Measuring cost savings directly induced by demand forecasts would require forecasts to be used throughout the planning process unchanged, and an experimental setting that excludes any other effect on the forecast accuracy, which remains an open research task.

Apart from direct financial gains, which according to practitioners in AI projects to date for the reasons just mentioned should not be the primary goal [11], more order deliveries have met the requested delivery date with the demand planning solution, thanks to material management that is in line with customer demand. Another significant benefit comes from automating parts of the planning process: Manual efforts are reduced because demand forecasts can be fed into the next planning step automatically.

However, for this to happen, the planners need to trust in the forecasts generated by AI methods. To gain this trust, the demand forecasts are analyzed and evaluated weekly by the material planners. For this analyzing and evaluating task, several dashboards were integrated in the demand forecasting solution, which were developed based on interviews with the planners and extensive user testing. The planners have been trained how to adjust these dashboards, so they are able to extend and edit the evaluations as needed.

This frequent checking makes sure that the demand forecasts are used for a growing number of products, thus realizing the expected benefits. The fact that the planners use the evaluation functionality on a weekly basis, and have added further products to the demand forecasting solution, confirms the recommendation of frequent reviews of benefits realization in IT projects [12].

Finally, the planners reported to gain insights into demand trends thanks to visualization not only of demand forecasts for individual products but also for product groups, and the possibility to compare them.

5.2 Costs

Costs for demand forecasting projects arise mainly from two sources: infrastructure, i.e. software and hardware used, and human resources. The computational resources needed for neural networks to be trained and to generate forecasts depend on several parameters, e.g. the complexity of the neural networks, the number of time series to forecast and the number of observations for each time series. Whether servers on-premise or cloud services are used, in either case their part of total project costs is small in comparison to the human resources needed. This ratio is reversed once the project is finished, the expected decision support and level of automation offered by the demand forecasting software are reached, and the AI solution is set productive.

For the demand forecasting solution that was developed, which serves to forecast weekly demand of approx. 700 products with an entirely automated workflow for 52 weeks using recurrent neural networks and public cloud-services, the infrastructure costs amounted to several hundred Euro per month, including quarterly retraining of the RNN models. The human resource costs amounted to approximately 30 man-months, comprising a full-time employed machine-learning engineer, a data scientist, a process expert and a material planner who worked part-time on developing the demand planning solution.

6 Success Factors of AI Projects for Demand Forecasting

6.1 Components of Project Success: Four Sides of the "Magic Square"

From interviews with the supply chain managers (i.e. those responsible for the stock levels, delivery reliability towards customers and bearing the costs for the AI project) as well as with project team members, and from project reviews during and after completion of the project, the success of this AI project depended on the following four components: 1. business and domain knowledge, 2. data, 3. AI methods and 4. user experience. For illustration purposes, they can be assembled to make up a "magic square" (Fig. 1):

Fig. 1. Components of project success: The magic square for AI projects

In the following, these four components are described in further detail.

A) Business and domain knowledge

The business typically defines the problem setting and the requirements for an AI-based solution. The problem definition includes e.g. the forecast horizon, the planning level, the knowledge about the sales dynamics and potential external drivers. The requirements can be expressed in terms of performance (e.g. accuracy or other business KPIs), interpretability or simplicity of the AI solution.

B) Data

Model input data should cover and reflect the aspects of the problem setting. It serves as basis for model development and refinement. Therefore, it should be a representative sample that allows to describe and model the demand dynamics.

C) AI methods

A growing number of AI and machine learning algorithms can potentially be applied to solve the problem. Here especially RNN as outlined in Sect. 2 for the modeling and forecasting of time series are of interest. To augment the model building process by incorporating prior knowledge (see Sect. 4) has given more accurate forecasting results. Learning from data is only one part of this process. More complex neural network architectures that e.g. include error correction terms or additional target variables are examples of this model building philosophy [5]. Remarkably, such a joint model building framework does not only provide superior forecasts, but also a deeper understanding of the underlying dynamical system. On this basis it is also possible to analyze and quantify the uncertainty of the predictions, as outlined in Sect. 4. This is especially important for the development of decision support systems as in the case of demand planning.

D) User experience (UX)

This side of the square considers the application of and work with the AI solution. Which role in the planning process is supported, i. e. who are the users of the AI solution? Which benefits does the user expect? How can the AI results be interpreted and visualized such that the user is willing to invest the needed time for this task and sees benefits in it? How is the AI solution maintained and serviced?

These four sides of the square are not independent from each other and must be balanced in order to implement a successful project. For instance, if the business department requires a forecast of the demand figures with an accuracy of at least 90% (i.e. 10% forecast error), a data sample is needed that covers all fluctuations (resp. patterns) of the demand dynamics as a basis for model building. The data sample must in turn be processed with methods (such as RNN) that are capable of memorizing and generalizing the demand patterns. Regarding user experience, users must be able to interpret and understand the results of the models.

The requirements of each side set constraints and condition for the others. For example, if one requirement of the business is that the models are interpretable, some model techniques might be only applied with restrictions (e.g. neural networks are often referred to as black boxes, whereas the results of decision trees or linear models are more transparent). It is also important to note that the sampling of the data should not be guided by mathematical-statistical principles alone. Of course, it is important to have a representative sample for the model building and statistically every data point has the same importance. However, a data point from a period which does not reflect the demand dynamics in the eyes of the business (e.g. observed in the financial crisis) should not be considered for the modeling.

The results of case study suggest that for project success it is indispensable that **experts representing each side** of the magic square work closely together in the project execution. Project work following **agile software development principles** appeared as another success factor as it allowed for changing priorities, to incorporate evolving requirements, and limit costs. The mentioned expert roles and agile proceeding will be described in greater detail in the following.

A) To provide a common understanding of the problem to be solved, explain the business process as-is, and consider dependencies to adjunct business processes, a **process expert** is needed on the project team. He or she knows the business domain and is able to define the needed sample and the relevant data attributes. In addition, this role has insights into factors that affect the demand figures, such as announced product market entries or end-of-life dates. With know-how about the entire sequence of planning steps, this role helps to make sure that the AI solution does improve planning: either AI can be integrated into the existing planning process, or it replaces some steps.

Historical data, particularly the selected sample including relevant attributes is the foundation on which AI can be applied to generate predictions. In today's business environments, there are usually many software applications in use which store sales data. It requires expert know-how to merge these different data sources and set up a data pipeline that provides updated input for each newly created prediction. That's what the so-called **machine-learning engineer** does. For development, he or she quickly sets up an experimental software environment, and after successful evaluation, implements a scalable software architecture to support Predictions generation and model tuning as needed.

To choose wisely from the universe of available AI methods and then apply the chosen algorithms, a **data scientist** is on the team. This role entails a detailed understanding of the mathematical models and available Deep Learning technologies, and ideally strives to uncover the true pain points of the planning personnel and the existing planning process.

Planners, or those employees using the Predictions in their planning tasks, are needed not only to make sure the developed solution provides a satisfying user experience, but also to iteratively evaluate the Prediction results, by visualizing and comparing them to the human planning figures used, and thus confirm the AI model and model parameters to be used.

When developing an AI demand prediction solution, a **UX designer** investigates the needs of the users, which are not always obvious, and makes sure the solution benefits them. He or she observes and interviews the planners before and after predictions are available to them, helps them visualize and evaluate the predictions, and supports the team in focusing on the most pressing pain points to be solved, e.g. which products and which Forecast horizon are of greatest interest to the planners, and how can they make the best use of the Forecasts.

Introducing the Predictions into the planning process signifies changes. Manual steps will be automated, human judgement will be enhanced or replaced by the Predictions, and both affect the planners' work routines. From our experience, it is essential that there is a "**driver**" on the team who works closely with the planners, to explain the changes and consistently make sure the Predictions are used, instead of recurring to the prior situation. This role can be taken on additionally by one of the roles above. Alternatively, the **product owner** for the developed solution takes on this responsibility.

With these experts on the team, an AI-based demand planning solution was successfully implemented, as depicted in Fig. 2.

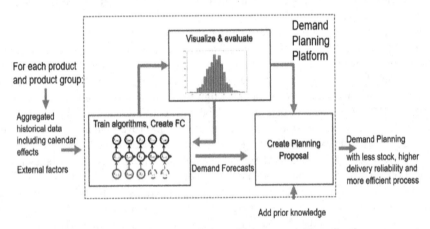

Fig. 2. Demand planning solution with integrated AI application

6.2   Agile Proceeding

In the studied demand forecasting project, requirements crept in, i.e. new requirements were added to the backlog of desired functionality, priorities changed, and the initial customer was replaced by someone who brought in a different perspective and different pain points to be addressed. As one insight from the case study interviews and confirmed by many software developing organizations [13], applying agile software development principles and techniques is an effective way to overcome the mentioned challenges. Primarily, assigning the roles of product owner, scrum master and development team helps clarify responsibilities. Working with a product backlog that can be extended at any point in time, that is groomed and prioritized by the product owner, and from which a limited number of requirements is selected for development in the next short development sprint (i.e., a two to three week-interval), have allowed for changing priorities and to always focus development efforts.

For efficient and productive teamwork, daily standups were done, which were used to communicate and clarify open issues at least once a day. Also, team retrospectives led to enhanced teamwork, as they offered a chance for each team member to reflect on the past sprint and thus allow the team to improve their collaboration. Possibly the most beneficial aspect was customer involvement: In developing the demand planning solution, the customer acted as product owner, thus participating in daily standups and responsible for prioritizing requirements with the team. In addition, the planners regularly evaluated the demand forecasts and gave valuable end user feedback. This aspect has been identified as critical success factor in software development projects [14] and is particularly important when developing innovative products using AI [15], such as the demand planning solution at hand.

7 Conclusion and Outlook

The case study presented an AI project that was carried out to establish a demand planning solution for supply chain management, particularly material planning. The aim of the solution was to improve the quality and efficiency of the planning process. The technological basis for the demand forecasts was a time-delay recurrent neural network (RNN) that is applied within the context of an automated modelling process. Automatically generated RNN models proved to be useful to predict the demand for approx. 700 products, particularly thanks to weekly evaluations of these predictions by the planners. The demand forecasts were transformed into planning proposals by combining the forecast with a measure of uncertainty. The forecast uncertainty was derived from the residual error distribution of each individual model. Analyses of the project results, interviews and project reviews, showed the following key success factors: bringing together data, AI methods, business and domain knowledge and user experience, each represented by an expert on the project team, as well as applying agile software development practices, i.e. following a SCRUM process.

Automated model building and so-called data citizenship is currently one major trend in AI and related research. Data science is a scare and costly resource. Hence, automated model building helps to overcome capacity bottlenecks of data science departments and enables the business departments to directly approach problem settings from the field of data science. Another trend in AI is towards data efficient models and training algorithms. Deep learning typically needs large data sets to fully exploit its potential. In contrast, the data volume in many industrial applications is rather small, data collection is costly and handling big data requires IT resources, data engineering knowledge and other administrative efforts like e.g. data security. Especially in demand forecasting, models for new products must be built based on rather small data sets. Here data-efficient algorithms or strategies like transfer learning come into play. Work currently in progress to extend and improve the demand planning platform addresses these two research fields.

References

1. Foellmer, H.: Alles richtig und trotzdem falsch? Anmerkungen zur Finanzkrise und Finanz-mathematik. MDMV **17**, 148–154 (2009)

2. McNeil, A., Frey, R., Embrechts, P.: Quantitative Risk Management: Concepts, Techniques and Tools. Princeton University Press, Princeton (2005)

3. Haykin, S.: Neural Networks and Learning Machines, 3rd edn. Prentice Hall, Upper Saddle River (2008)

4. Zimmermann, H.G., Grothmann, R., Tietz, C., Jouanne-Diedrich, H.: Market modeling, forecasting and risk analysis with historical consistent neural networks. In: Hu, B., et al. (eds.) Operations Research Proceedings 2010. Springer, Heidelberg (2011). https://doi.org/10.1007/978-3-642-20009-0_84

5. Zimmermann, H.-G., Tietz, C., Grothmann, R.: Forecasting with recurrent neural networks: 12 tricks. In: Montavon, G., Orr, G.B., Müller, K.-R. (eds.) Neural Networks: Tricks of the Trade. LNCS, vol. 7700, pp. 687–707. Springer, Heidelberg (2012). https://doi.org/10.1007/978-3-642-35289-8_37

6. Runkler, T.A., Grothmann, R., Bamberger, J.: Optimierung industrieller Logistikprozesse mit Verfahren der Schwarmintelligenz und rekurrenten neuronalen Netzen. KI Künstl. Intell. **24**(2), 149–152 (2010)

7. Crone, S.: Neuronale Netze zur Prognose und Disposition im Handel. Betriebswirtschaftlicher Verlag Gabler, Wiesbaden (2009)

8. Gleißner, H., Femerling, J.C.: Logistik: Grundlagen – Übungen – Fallbeispiele. Gabler Verlag, Wiesbaden (2007)

9. Wei, W.S.: Time Series Analysis: Univariate and Multivariate Methods. Addison-Wesley Publishing Company, New York (1990)

10. Costello, K.: Top 3 benefits of AI projects. https://www.gartner.com/smarterwithgartner/top-3-benefits-of-ai-projects/. Accessed 20 Aug 2020

11. Van der Meulen, R., McCall, T.: Nearly half of CIOs are planning to deploy artificial intelligence. https://www.gartner.com/en/newsroom/press-releases/2018-02-13-gartner-says-nearly-half-of-cios-are-planning-to-deploy-artificial-intelligence. Accessed 22 Aug 2020

12. Braun, J., Mohan, K., Ahlemann, F.: Realising value from projects: a performance-based analysis of determinants of successful realisation of project benefits. Int. J. Project Organ. Manag. **8**(1), 1–23 (2016)

13. 14th Annual State of Agile Survey Report, VersionOne (2020). https://stateofagile.com/#ufh-i-615706098-14th-annual-state-of-agile-report/7027494. Accessed 27 Aug 2020

14. Cerpa, N., Bardeen, M., Kitchenham, B., Verner, J.: Evaluating logistic regression models to estimate software project outcomes. Inf. Softw. Technol. **52**(9), 934–944 (2010)

15. Abramov, J.: An agile framework for AI projects. https://towardsdatascience.com/an-agile-framework-for-ai-projects-6b5a1bb41ce4. Accessed 27 Aug 2020

Author Index

Bohanec, Marko   30
Boshkoska, Biljana Mileva   44

de Almeida, Adiel Teixeira   18
Delibašić, Boris   110
Dowie, Ulrike   124

Eriksson, Kristina   84

Grothmann, Ralph   124

Hendberg, Ted   84
Huang, He   3

Kikaj, Adem   30

Lebeau, Philippe   3
Levashova, Tatiana   97
Liu, Jinyi   69

Macharis, Cathy   3
Mommens, Koen   3

Ponomarev, Andrew   97

Radovanović, Sandro   110
Rančić, Sanja   110
Roselli, Lucia Reis Peixoto   18

Shilov, Nikolay   97
Smirnov, Alexander   97
Stacey, Patrick   69
Stipeč, Anton   44

Wieland, Georg   55

Zeiner, Herwig   55

Printed in the United States
by Baker & Taylor Publisher Services